The Physics of Optical Recording

Kurt Schwartz

The Physics
of Optical Recording

With 81 Figures and 11 Mostly Coloured Plates

Springer-Verlag

Berlin Heidelberg New York
London Paris Tokyo
Hong Kong Barcelona
Budapest

Professor Dr. Kurt Schwartz (Shvarts)

Physics Institute, Latvian Academy of Sciences, Riga,
Salaspils 1, LV-2169 Latvia

Present address: Physical Institute, University of Heidelberg,
Philosophenweg 12, D-69120 Heidelberg, Germany

ISBN 3-540-52237-9   Springer-Verlag Berlin Heidelberg New York
ISBN 0-387-52237-9   Springer-Verlag New York Berlin Heidelberg

Library of Congress Cataloging-in-Publication Data. Shvarts, K. K. (Kurt Kurtovich) The physics of optical recording/ Kurt Schwartz. p. cm. Includes bibliographical references and indexes. ISBN 3-540-52237-9. – ISBN 0-387-52237-9 (U.S.) 1. Optical data processing. 2. Laser recording. 3. Optical storage devices. I. Title. TA1635.S48   1993 621.39'767–dc20   93-14006

Typesetting: Camera-ready copy prepared by the author using Microsoft® Word 5.0

56/3140-5 4 3 2 1 0 – Printed on acid-free paper

Preface

Optics and optical recording have a long history. Already in the ancient temples of Egypt and in Stonehenge, optical imaging was used in religious ceremonies. As early as the 17th century, optical instruments such as the telescope and the microscope played a crucial role in the development of the natural sciences, which, in turn, altered humanity's perception of itself and the universe. Nowadays, the science of optics is not only a means of gathering information but, also, of information transfer, analysis and processing. Our reliance on optical computing has become crucial since 1982, when compact discs (CD's) first became widely available. Optical memory discs, CD ROM (read-only-memory), WORM (write-once-read many) and erasable read-write discs are now produced industrially. However, WORM and erasable discs comprise only a small field of optical imaging and information processing. Recent research into optical recording introduced new horizons for optical computing and real time information processing, such as, three-dimensional holographic memory, time-domain holography, nonlinear optical filtering, etc. The development of these unique possibilities is conceivable only with new recording materials and novel recording processes. For such progress, one should be familiar with the "pen and paper" of optical recording – the laser and the optical medium. Thus, this book will be a guide to novel applications of optical recording materials. The reader will find a description of the principles of optical recording and optical processing, which allow him to find novel applications for information storage and optical imaging.

Chapter 1 begins with a brief historical review of optics and optical recording, followed by a synopsis of the present state of optical processing and memory systems. In Chap. 2, the principles of real time optical recording (digital and holographic) are illustrated. Special attention is payed to holographic recording, and the parameters of two and three-dimensional holograms, with their specific recording capacities, are discussed. Chapter 3 deals with lasers (with special attention devoted to diode lasers) and nonlinear optics. This section also contains a description of light modulators and photoelectric detectors indispensable in any equipment for optical recording and readout. In Chap. 4, inorganic and organic recording materials and their photosensitive properties are presented. New materials and recording processes and time-domain holography for three-dimensional holographic memory are shown, which open new possibilities for mass storage and associative optical memory systems. This is followed (Chap. 5) by a detailed discussion of photothermal recording processes (their advantages and limitations) commonly used to produce optical memory discs. So, the topic of this book is the physics of optical recording, including the equipment required for photoinduced process studies. The technology of industrially produced memory discs and the digital optical recording technique is beyond the scope of this book. These problems are discussed in detail by Alan B. Marchant *Optical Recording – A Technical Overview* (Addison-Wesley, Reading, Massachusetts 1990).

This book is intended for professionals in optical recording and information processing as well as for scientists and engineers involved in optics and optoelectronics. For professionals in optical recording materials, the most informative parts of the book are Chaps. 4 and 5. Newcomers to the field might find the first section, namely: Chaps. 1–5, a helpful guide for R & D into optical recording and processing. Furthermore, the book may be useful for undergraduate and graduate students in optics and optoelectronics.

The book was completed during my stay at the Physical Institute of the University of Heidelberg. I want to express my deep thanks to Prof. Dr. Gisbert zu Putlitz, who made this stay possible, for his assistance and for many fruitful discussions.

I would like to express my gradiptude to Dr. A. Ozols, Dr. N. Aristov, Dr. S. Noehte and Prof. E. Silinsh for helpful comments and discussions during the preparation of the manuscript. Thanks are also due to U. Labucis for the photographs (Plates 6–9) and to Kurt Fridrichson for the pictures (Plates 4 and 5), which demonstrate the imagination of our visual impressions.

Heidelberg, November 1993   *K. Schwartz*

Contents

1. Introduction

> The mirror reflects all objects, without being sullied.
> *Confucius* (551–479 B.C.)

Man begins with the word and the written symbol (ideogram) which are the main expressions of human thought. The way from the abstraction in the rock painting (30.000 years ago!) to optical computing was long. However, our visual impressions were important from ancient times up to present days, where they play an important role in the modern computer technique, particularly in visual display units and picture analysis.

1.1 Visual Impressions and Optics

Optical imaging by the eye is the main source of information we receive from the world around us. Up to 90% of the total information registered by our sensory receptors is obtained through our vision. We look at beautiful landscapes, recognize our friends, read books, determine the weather, enjoy the sight of stars and explore the universe. However, in terms of data processing rate, storage capacity, and even in terms of image quality, computer based optical imaging and recording systems are becoming, or already are, competitive with human vision.

Nowadays, human vision and optical processing are investigated by optics - one of the oldest sciences. From ancient times, optics and geometry have had a great influence on astronomy, navigation and metrology. Euclid (365–300 B.C.), the founder of geometry, based his concept of the line on the path taken by a light beam. In all ancient civilizations, mirrors and lenses – the simplest optical instruments – were known (a description of light focussing with lenses was given by Aristophanes (445–383 B.C.) in the comedy "The Clouds").

The influence of optics on the natural sciences rapidly increased during the Renaissance after the invention of the microscope and the telescope. In the 19th century, the discovery of optical spectroscopy, photoelectric and electro-optic effects stimulated the development of the fundamental theories of modern physics such as quantum mechanics and Einstein's theory of general relativity and provided the basis for the technology of optoelectronics. The most important inventions and discoveries in optics are given in Table 1.1. which illustrates the long way to real time optical recording.

Table 1.1. A brief view of the history of optics

Phenomenon	Inventor	Year
Corpuscular theory of light	Pythagoras Empedocles	VI-V B.C.
Light beam; directed light propagation	Euclides	IV B.C.
Laws of reflection	Archimedes	III B.C.
First optical instruments – microscope and telescope	I. and Z. Janszen, H. Lippershey	1590–1608 A.D.
Microscope with ocular and objective; optimization of the telescope for astronomical observation	G. Galilei	1609
Laws of reflection and refraction	W. Snell (van Roijen) R. Descartes	1637
Light diffraction	F.M. Grimaldi	1665
Application of the microscope to physics, chemistry and biology	A. van Leeuwenhock, R. Hooke	about 1665
Corpuscular theory of light; dispersion; reflecting telescope	I. Newton	1668
Double refraction in $CaCO_3$ crystals	E. Bartholinus	1669
Light interference	T. Young	1801
Light propagation in anisotropic media; principles of crystal optics	A.J. Fresnel	1815
Photochemical action of light	T. Grotthus	1817
Photoconductivity of selenium	J.W. Hittorf	1852
Electromagnetic wave theory	J.C. Maxwell	1865
Photoelectric effect (electron emission from metals)	H. Hertz	1887
Discovery of electromagnetic waves	H. Hertz	1887
Thermal emission and photon theory of electromagnetic radiation	M. Planck	1900
Photon theory of photoelectric effect and photoinduced processes	A. Einstein	1905
Nonlinear light absorption	S. Wavilow	1925
Phase-contrast method of optical imaging	F. Zernike	1934

Table 1.1 (continued)

Phenomenon	Inventor	Year
Invention of holography	D. Gabor	1948
First laser Al_2O_3-Cr	T. Maiman	1960
Invention of the He-Ne gas laser	A. Javan	1960
Second harmonic generation in quartz (SiO_2) crystals	P. Franken	1961
Three dimensional holography	Yu.N. Denisyuk	1962
Holographic information storage [1.1]	P.J. Van Heerden	1963
Mode synchronization and ultrashort laser pulse generation: He-Ne laser (2.5 ns) [1.2] Nd-YAG (40 ps) [1.3] Dye lasers (5 ps) [1.4] Dye lasers (6 fs) [1.5]	L.E. Hargrove et al. M.Di. Domenico et al D.J. Bradley et al. R.L. Fork et al.	1964 1966 1968 1987

1.2 Imaging in the Eye and Optical Recording

The human eye is similar to a camera with a lens that projects an image onto the retina, playing the role of the film. However, our visual impressions are much more complicated than the imaging on a film. Such a sensory image begins with biochemical recording in the retina by the cone-shaped (color vision) and rodlike cells (dark or night vision). After a complex biophysical information process, a "neural" image is created in the eye and transferred by the neuron fibers to the brain. Our visual impression is a real time recording process. The relaxation time of the neural imaging, i.e., the delay between the light absorption in the retina and the "neural" image formation, is approximately 0.05 seconds for night vision and 0.02 seconds for color vision. Therefore, the imaging process in the human eye is much slower than in a digital optical computer (about 0.1 µs by photothermal recording in optical memory discs). However, the large number of photoreceptors in the retina (approximately 7 millions for color and 130 millions for the dark vision) and the ability of parallel information processing (made possible by the one million neuron fibers connecting the retina to the brain) ensure an extremely high data processing rate in our visual perception system: about one hundred million signals per second! This is a much higher signal rate than those of other human senses and comparable with present digital computers and optical communication (wave guide) systems [1.6, 7, 8].

On the other hand, the stored information density in optical memory discs (up to 10^8 bit/cm^2, Plates 1–3) is much higher than the color vision sensor density in the retina (about 10^6 cones/cm^2). Furthermore, the storage time of the data, i.e. the number of seconds or years that the medium retains the data, is guaranteed in the case of the disc, but completely random in the human brain. However, in other instances, the eye-brain system is more effective than man-made optical data retrieval and storage devices. For example, the eye is one of the most sensitive photodetectors known: the dark-adapted eye (rod-like cells) can detect approximately 2 to 10 light quanta, which is much higher than that of common technical photodetectors.

In addition, the eye is not only a biophysical sensor (recording medium), but, also, an optical data processor with multifunctional pattern analysis. The neural image contains much more information than the initial optical image on the retina. For example, we recognize persons, things or details in an image with the aid of the information processing of color, contrast and pattern details via the photoreceptors. Thus, the final, much more generalized, or abstracted, image is formed in our brain and provides us with information on the surrounding world that has been more thouroughly evaluated than the best computers can manage.

Table 1.2 summarizes such comparisons between human and machine capabilities. The computer is evidently superior as far as access time and pure numerical analysis is concerned. However, the "human brain" computer has a much larger storage capacity and greater capability for complex tasks, such as

Table 1.2 Data processing in computers and the human brain [1.7]

Parameter	Digital Computer	Brain
Storage capacity [Gbit]	200	1000
Access time [s]	$10^{-9} - 10^{-6}$	$10^{-4} - 10^{-3}$
Numerical analysis	very good	poor
Pattern recognition and analysis	poor	very good
Self-organizing and learning	poor	excellent

pattern recognition and analysis, self-organization, and learning. Our visual experience with the world is much more than a simple reflection or reproduction – it is a mental image, an abstraction, a picture enriched with feelings and sensations (Plates 4, 5). The future development of optical memory devices must combine the technical advantages with the associative processing of the human brain.

1.3 Digital and Holographic Recording

Direct, or real time optical imaging required the invention of laser techniques as well as, of course, new recording media to hold the images. In principle, these media are simply networks of optical switches which are turned on or off when they are irradiated by photons at the proper wavelength. The switching action can be a light induced change in color, reflectivity, transmissivity, magnetization, or even a phase change such as local crystallization, as seen in semiconductor films on Plates 1–3.

Today, real time optical recording is usually binary (digital), e.g., a recorded spot means "1" and an empty place means "0". New possibilities for optical memory systems were opened by holography which was invented by Denis Gabor in 1947 [1.10]. Holography is a general recording method by coherent waves (light, acoustic, etc.). It can be used for any information recording – from digital to pictorial and volume imaging. The fields of application of optical holography are diverse, e.g., holographic optical elements (HOE) - lenses, diffraction gratings, coherent filters etc.. Holographic filtering can be used for pattern recognition and nonlinear information processing which is necessary for associative memory systems [1.11, 12].

Some of the optical processes occurring on the retina are similar to those occuring in a spatial holographic filter, which, once more, demonstrates the complex functions of the eye as an optical processor [1.13]. Sometimes, such processing leads to optical illusions: looking at *Kanizsa* triangles, the illusion of black and white contours is created in the mind (Plate 10). The mechanism of this illusory perception has been explained by *Ginsburg* [1.14]. Plate 11 shows an example of the artistic use of optical illusion [1.15, 16].

Holography is also the basis of a new generation of associative optical memory systems. Such "neural optical computers" combine the principles of human learning in the brain with optoelectronic data processing [1.17, 18].

1.4 Recording Time and Data Rate

As has been prewviously mentioned, our visual perception is not very fast, with a relaxation time of approximately 0.02–0.05 seconds. This limitation is exploited in motion pictures where the typical projection rate is 30 images (frames) per second (cinema, TV).

If the image rate is slowed down even further, one has the beautiful and weird stroboscopic effect used abundantly in discotheques and in the circus (Plate 9). The stroboscopic effect was first used by Michael Faraday, and somewhat later, by Joseph Plateau in 1836 in the study of vibrating and rotating objects. This is now a common technical application [1.19].

The real time optical recording takes place over a much shorter time than optical imaging in the eye. However, the digital photothermal recording in optical memory discs (see Plate 1) has a time limit of 10–50 ns caused by the conductivity of heat in the illuminated spot. This gives the maximum data rate of 10^8 bits/s. For faster data processing new recording media must be developed by shorter laser pulse excitation. However, this lack of data rate can be avoided by performing a faster recording using a high intensity laser pulse of shorter duration or by the use of a parallel multichannel recording-readout system (similar to human vision). Note that the recording time in every case is limiting the beginning of the readout which is possible only after the complete finish of the recording.

A new field of fast optical imaging, the so-called "light in flight capturing", was developed by *Abramson* and *Valdmanis* [1.20], i.e., short pulse holographic imaging in the femto-second region. The kinetics of light scattering with the spatial distribution function in the femto-second time scale was studied in this method.

The absolute limit imposed on the resolution time (i.e., the shortest possible time of measurement) for any physical process is dictated by the uncertainty relation:

$$\Delta E \cdot \Delta t \geq h \ . \tag{1.1}$$

For the visible spectral region ($\Delta E \approx 2$ eV), this gives a minimum time resolution of $\Delta t = 3 \cdot 10^{-16}$ s. This is close to the length of the shortest generated laser pulses (see Table 1.1). In optical memory discs the recording time is much longer (about 100 ns) than the resolution time. Such long recording times are caused by the nature of the photothermal recording (Chap. 5) and limit the data rate in optical memory discs up to 100 Mbit/s. Photodetectors, operating by the photoelectric effect, have a data rate of the order of 20 Gbit/s. This corresponds to a possible shorter recording time of 50 ps. At present, however, there are no optical recording media available with such a fast relaxation. Moreover, it is quite difficult to get such short intense light pulses from laser diodes which are commonly used in memory devices.

This large discrepancy between optoelectronic and photothermal media stimulated the search for light sensitive materials with shorter recording times in the picosecond range. It is not easy to combine high light sensitivity, short recording time and long storage time in a single material. One such species is the bistable semiconductor, which may become particularly important in the development of picosecond microprocessors in "pure" optical computers [1.21].

1.5 Storage Capacity

The data optical storage capacity is limited by the properties of the medium and the optical system (setup). High quality optical systems are necessary with minimal aberrations for optical recording (Plates 7 and 8). The diffraction limits the minimal size of the recorded spot ($d_{min} \geq \lambda$, where λ is the recording or readout wavelength). This gives the maximum storage density in a two-dimensional medium of $1/\lambda^2 \approx 5 \cdot 10^8$ bit/cm^2, and, in a three-dimensional medium of $1/\lambda^3 \approx 10^{13}$ bit/cm^3 (for the wavelength $\lambda = 450$ nm). But the complications of three-dimensional optical recording and readout make the two-dimensional systems more common.

It is enticing to avoid the diffraction limit of the optical storage, i.e., to get from one diffraction limiting spot more information than one bit. Such a possibility was introduced by the Persistent Spectral Hole Burning (PSHB) method or phenomena. This method is based on high resolution spectroscopy with zero phonon lines [1.22, 23]. Color centers and impurity defects in crystals, organic materials, dyes (especially dye solutions in polymers) are available for PSHB. The recording leads to light induced absorption change, i.e., to narrow spectral bands (the linewidth of the spectral hole is much smaller than the bandwith of the absorption spectrum). Therefore, at one spot, up to 10^3 bits can be recorded. This gives a total storage density of 10^{11} bits/cm^2. For high resolution spectroscopy the readout must be performed at liquid helium temperatures, which is more expensive than the data storage in discs with magneto-optical or phase change recording.

1.6 Applications of Real Time Optical Recording

Real time optical recording is switching with light , i.e., a light induced change of the optical, or other properties, of the medium. Generally, light induced switching is reversible and can be erased by light or another physical mechanism. For optical processing, the recording and readout times must be of the order of nanoseconds and the storage time as long as possible (years). The storage density and light sensitivity have additionally to be as high as possible.

The present state of digital optical recording in optical memory discs has reached the diffraction limit of 10^8 bits/cm^2. These are commercially available as CD ROM, WORM and erasable discs [1.8]. However, the light sensitivity of optical memory discs (approximately 1 nJ/bit) is 10^4 times lower than that for high resolution photographic films. The high light sensitivity of common photographic films is reached by the amplification of the primary light induced effect during the chemical development. Such an amplification of photoinduced effects usually is absent in real time recording materials (some amplification in ferroelectric materials can be received in an external field, see Chap. 4). Nevertheless, optical memory discs are widely used for digital optical recording and the recording technique, as well as the technology of optical disc production, is well known [1.8, 24, 25]. In commercial optical memory discs two kinds of materials (magneto-optic and phase change) are used and the recording parameters achieved are practically the physical limit for this class of materials.

One more limitation on commercial optical memory discs is the digital recording; still, for many applications, analogue and holographic recording is necessary. On the whole, any optical recording and optical processing needs some specific real time optical recording medium. This is possible only by understanding the recording processes in the medium and using a special preparation technique.

The research and development of new real time optical recording materials is a complex problem depending on the field of application. Special treatment is devoted to holographic (associative memory, pattern recognition, holographic optical elements) and fast (picosecond) real time optical recording. Direct optical recording, as well as computer generated holograms recorded by light or electron beams are important applications. Such computer generated or synthetic holograms can be produced without, or with, minimal aberrations.

This book will be predominantly concerned with the origin of optical recording. Chapters 2 and 3 will provide the principles and techniques of digital and holographic recording. Still, the prevalent interest are the photoinduced processes and mechanisms in recording media (Chaps. 4 and 5). Such analyses could be relevant for further activities in this field.

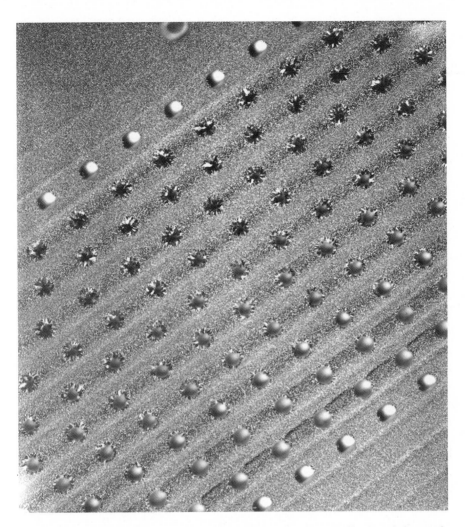

Plate 1. A transmission electron microscope photograph of optically recorded spots in $Te_{0.8}Se_{0.1}Sb_{0.1}$ films (thickness 100 nm; the size of a single spot is approximately 1 μm). The recording process takes place under a short laser pulse (approximately 100 ns) and leads to a local amorphization of the polycrystalline film (the sharp spots in the top and bottom tracks). The optical erasure is a local recrystallization of the amorphous spots which is induced by a longer laser pulse (10 μs and more) with lower light intensity. Erased tracks with different exposure are shown between the sharp top and bottom tracks. From [1.9]

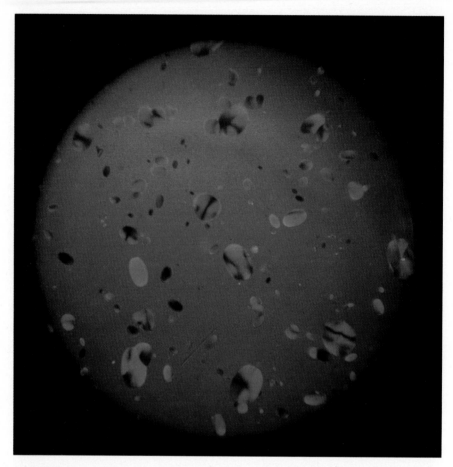

Plate 2. Photoinduced crystallization in antimony chalcogenide films: spheroidal crystallites of $Sb_{0.3}S_{0.7}$ as seen through a polarizing microscope. The size of the largest crystallites is approximately 10 μm

Plate 3. Photoinduced needle-crystallites in $As_{0.1}Sb_{0.3}S_{0.6}$ films through a polarizing microscope (the magnification is the same as in Plate 2). Similar processes (crystallization, amorphization) in antimony chalcogenide films are used for optical recording

Plate 4. The human brain conceives the optical image with abstraction, feeling and sense. Landscape-painting "Romantic" by Kurt Fridrichson (Riga)

Plate 5. An abstract portrait-painting. "Made in..." by Kurt Fridrichson (Riga)

Plate 6. Light scattering by a holographic optical element. Holographic elements are used in optical memory devices instead of conventional optical elements (lenses, beam splitters etc.)

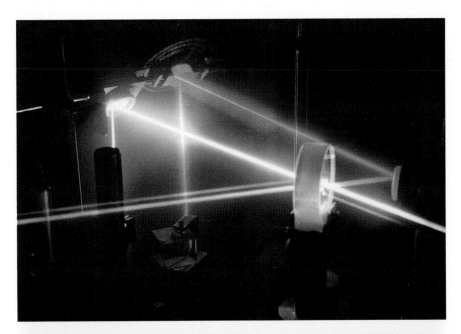

Plate 7. Light reflection and refraction in optical instruments

Plate 8. Spherical aberration of a mirror: the lateral rays have a shorter focus than the central reflected rays. A high spatial resolution and a high storage capacity in optical memory devices are possible only with high quality optics with minimal aberrations

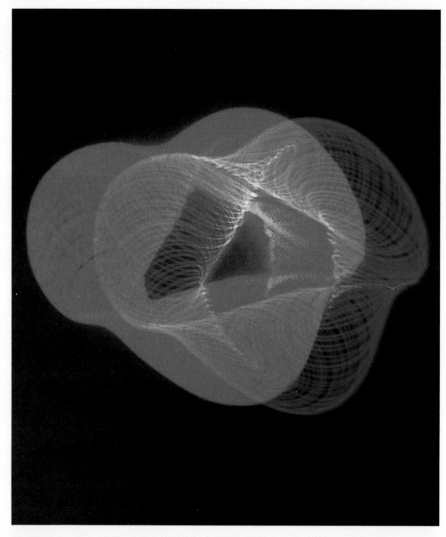

Plate 9. Red (He-Ne) and blue (argon) laser beam patterns produced by reflecting from two vibrating mirrors. The photographs show a superposition of optical Lissajous figures

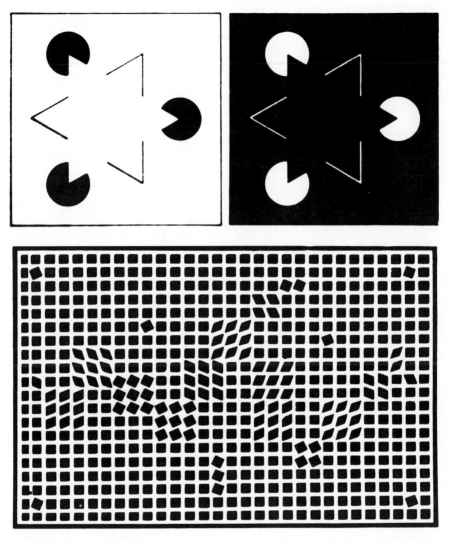

Plate 10. The Kanizsa triangles are geometric contrast illusions which are produced by spatial filtering in the human eye [1.26]

Plate 11. Optical illusions are used in op art. Painting by Victor Vasarely "Tlinko" showing an illusion of movement

2. The Principles of Optical Recording

Optical recording in solids is produced by light induced changes in the physico-chemical properties of a medium. Recording can be done on a real time scale without a later development process (unlike the case of silver halide photoemulsions where subsequent fixation is necessary). In this chapter, we discuss the basic principles of coherent (holographic) and noncoherent optical recording.

2.1 The Invention of Photography and Real Time Recording

The photochemical action of light on silver halides was observed by Schulze in 1727 [2.1]. More than a hundred years passed before Niepce, Daguerre and Talbot used this reaction to produce a photographic image. The main steps in

Table 2.1. The development of photochemistry and photography

Process	Author	Year
Photolysis of silver halides (AgCl, AgI)	J.H. Schulze	1727
Silhouette images on silver halide films	H. Devi	1802
	T.W. Wedwood	
The general law of photochemistry (only the absorbed light is photochemically active)	Th. Grotthus	1817
Reversible coloring of CaF_2 crystals under ultraviolet radiation	Th.J. Pearsall	1830
Chemical development of a latent image (with high contrast amplification)	J.N. Niepce,	1837
	L.J.M. Daguerre,	
	W.G.F. Talbot	
The reciprocity law:	W.F. Herschel	1842
the product of the photoinduced reaction is proportional	R.W. Bunsen	1855
to the exposure $(I \cdot t)$	H.E. Roscoe	1855
The quantum character of photoinduced processes: each photon induces one reaction	A. Einstein	1912

the development of the photochemistry related to photography are summarized in Table 2.1. The first reversible photochemical reaction, i.e., a potentially useful real time recording process was described by *Pearsal* in 1830 [2.2] and dealt with the behavior of CaF_2 crystals irradiated with ultraviolet light.

Before the photon theory of light was accepted, the efficiency of these photochemical processes was generalized by *Herschel, Bunsen* and *Roscoe* (Table 2.1) by the *reciprocity law*

$$Q = k_0 \cdot I \cdot t \ , \tag{2.1}$$

where Q is the concentration of the product of the photoinduced reaction, k_0 the reaction rate constant, I the light intensity and t the exposure time. Equation (2.1) is valid only for linear processes when k_0 is independent of I and t. Usually k_0 decreases after long exposure times (saturation) and is generally a function of both the intensity and time. At high light intensities, nonlinear processes dominate (Sect. 3.3).

Later, Einstein postulated the discrete character of light induced processes (e.g. the photoelectric effect, photochemical reactions, luminescence) and for a quantitative description of the light induced reactions proposed the quantum efficiency:

$$\eta_q = \frac{N}{N_0}, \tag{2.2}$$

where N is the number of product particles of a given photoinduced reaction (e.g., the electrons in photoelectron emission) and N_0 is the number of the absorbed photons. The limit $\eta_q \leq 1$ usually holds only for the primary light induced process. For example, in silver halide photomaterials η_q for the primary reaction of latent image formation is close to 1, but in the subsequent chemical processing, the photoreaction (i.e., silver atom generation) is amplified by 10^6 to 10^8 times (the upper limit is 10^9 and is reached only in special prepared films). The amplification depends on the light sensitivity and the resolution of the emulsion. For example, in photoemulsions with a high resolution of more than 2000 lines/mm, the amplification is smaller by between 10^5 and 10^6. The quantum efficiency of photoinduced reactions can be greater than unity by excitation of higher energy levels of the system (of atoms, molecules or the lattice of a solid), e.g., by irradiation with vacuum ultraviolet or X-ray photons [2.3].

Systematic investigations of photoinduced processes in solids first began in the 20th century. One of the groundbreaking experiments was done by *Röntgen* on the photoconductivity of colored NaCl crystals [2.4]. The first theoretical description of radiation induced processes in dielectrics was provided by *Pohl* and colleagues ([2.2], p. 364–366). They explored point defects in alkali halide crystals (doped, additively colored, irradiated) and luminescence in doped and irradiated crystals. Pohl also observed a transistor effect in colored alkali halides. This work was one of the foundations of modern solid state physics and also had a strong influence on advances in optical information recording.

Table 2.2. The Principle of optical recording processes

Light sensitive material	Process	Example	Year, references
Photochromic crystals or glasses	(1) Charge transfer (2) Orientation of dipoles (dichroism) (3) Phase transitions *Recording* : (i) Illumination and heating (ii) Illumination at (λ_1) and (λ_2)	Light induced optical anisotropy (dichroism) of M-centers in NaF and KCl crystals Photoinduced F-center coagulation in additively colored alkali halides	1967 [2.6] 1974 [2.2]
Photorefractive electro-optic crystals, optical ceramics	(1) Charge transfer (2) Photovoltaic effect *Recording*: (i) Illumination (ii) Illumination in an electric field	First phase holograms in electro-optic crystals $(LiNbO_3)$ Fixation and erasing of optical recording in $Sr_{0.75}Ba_{0.25}Nb_2O_6$ crystals by an external electric field Coherent selective erasure of superimposed holograms in $LiNbO_3$	1968 [2.7] 1969 [2.8] 1975 [2.9]
Amorphous semiconductors	(1) Charge transfer (bond breaking) (2) Phase transition *Recording*: (1) Illumination (2) Illumination and heating	Hologram storage in As_2S_3 films Optical memory discs (SbSeIn, SbIn etc.)	1971 [2.10] [2.35]
Magneto-optic semiconductors	(1) Phase transition (2) Charge transfer	Photoinduced thermal phase transitions in magneto-optic materials (MnBi etc.)	1968 [2.11]

Table 2.2. (continued)

Light sensitive material	Process	Example	Year, references
		Light induced magnetization change in $Y_3Fe^{3+}_{5-2x}(Fe^{2+}Si^{4+})_xO_{12}$	1967 [2.12]
	Recording: Illumination in a magnetic field		
Organic dyes, polymers	(1) Electron transfer (2) Proton transfer (3) Isomeric reactions (4) Polymerization	Photoinduced hydrogen transfer in organic dyes Optical and holographic recording in thermoplastics	1974 [2.13] 1963–1966 [2.14, 15]
	Recording: (i) Illumination (ii) Illumination in an electric field (iii) Illumination and heating		
Solids with zero-phonon lines (Hole burning spectroscopy)	Orientation of molecular dipoles	Optical recording on zero-phonon lines in porphyrazine styrol solution at 1.8 K	1974–1975 [2.16, 17]
	Recording: Illumination (by tunable narrow band) at $T \approx 1-2$ K		
Luminescent materials	Charge transfer (trapping and recombination)	Light induced recombination luminescence in doped CaS (recording at $\lambda_1 \approx 400$ nm; readout $\lambda_2 \approx 1000$ nm; output signal – luminescence at $\lambda_3 \approx 600$ nm)	1989 [2.18]
	Recording: Illumination		

The development of real time optical recording materials was only able to begin after the invention of lasers (Table 1.1). The basic concepts of optical recording were first discussed by *van Heerden* [2.5]. The main recording processes are presented in Table 2.2. Most of them lead to light induced charge transfer processes or phase transitions with a subsequent change of the optical constants, the index of refraction or absorption of the recording material.

2.2 Optical Recording Processes

Optical recording occurs through light induced changes in the property of the *recording medium*. When materials are illuminated, the incident photons can be absorbed, reflected (scattered) or transformed into other radiation (luminescence, inelastic scattering, photoelectric effect). For optical recording, only those photons which are absorbed are important, and, in particular those absorbed photons which induce an optical property change with a high quantum efficiency.

In general, every light induced change in the property of a material can be used to make an optical recording. However, for practical applications, usually a change in the light induced complex index of refraction, $\tilde{n}$, given by:

$$\tilde{n} = n - i\kappa , \tag{2.3}$$

where n is the index or refraction and κ the absorption index[1], or in the optical path length $d_{opt} = d \cdot n$, are acceptable. On the whole, a change in the complex refractive index always leads to the changes in both Δn and $\Delta\kappa$. However, often, in a particular material change, either Δn or $\Delta\kappa$ is larger. This can be illustrated by noting that when $\Delta\kappa$ dominates ($\Delta\kappa \gg \Delta n$), the medium is an amplitude recording material and in the other case ($\Delta n \gg \Delta\kappa$), a phase recording material. Phase recording occurs when the refractive index change dominates or when the profile of a transparent material is changed (i.e., the thickness of the materials is modulated). Phase materials have a small light absorption and are important for high efficiency holographic recording (see Sect. 4.2.3).

Phase recording materials were first discovered by *Ashkin* et al. [2.20] by a decrease in the modulation depth upon illumination in electro-optic modulators (LiNbO$_3$, LiTaO$_3$, and BaTiO$_3$ crystals). Similar effects were seen by *Peterson* et al. but not explained [2.21]. Further investigations of the disturbed modulation (i.e., optical damage [2.20]) in electro-optical crystals showed that the origin of this phenomenon was the light induced birefringence change

---

[1] The absorption constant κ is proportional to the linear absorption coefficient of light $k_a = 4\pi k/\lambda$ where λ is the wavelength of light in vacuum. The values of k_a and k are different for many orders of magnitude in various spectral regions (Sect. 3.1), but the refraction index n has in the whole spectral region a magnitude of 1–4 [2.19]. However, the amplitude or phase recording is determinated only by the magnitudes of Δn and $\Delta\kappa$.

Table 2.3. The discovery of electro-optic effects and photorefraction in dielectric materials

Phenomenon	Author	Year
Change of the refractive index of glasses and liquids (1880) in an external electric field; electro-optic Kerr effect[a] [2.23, 24]	J. Kerr	1875, 1880
Linear electro-optic effect in quartz crystals (SiO_2) [2.25, 26]	W.C. Röntgen A. Kundt	1883
General theory of electro-optic effects in crystals [2.27]	F. Pockels	1894
Anomalous photovoltaic effect in ferroelectrics [2.28]	A.G. Chynoweth	1956
Laser induced Kerr effect in liquids [2.29]	G. Mayer, F. Gires	1964
Photorefraction in electro-optic crystals ($LiNbO_3$, $LiTaO_3$, $BaTiO_3$) [2.20]	A. Ashkin et al.	1966
Correlation between photorefraction and the anomalous photovoltaic effect in $LiNbO_3$ and $LiTaO_3$ crystals [2.30]	A.M. Glass, D. von der Linde, T.J. Negran	1974

[a] Similar investigations of the quadratic electro-optic effect (i.e., Kerr effect) were carried out by *G. Quinke* (1880) and *H. Brongersma* (1882) in glasses

[2.22]. Such change of the refractive index or photorefraction effect[2] in electro-optical crystals leads to light induced intrinsic electrical fields (Table 2.3). A more detailed discussion of the photorefraction mechanism is given in Sect. 4.2.1).

2.3 Survey of Optical Recording Media and Their Uses

The first optical recording material used for imformation storage in libraries and archives was a silver halide film. Using high resolution microphotography *Goldberg,* in 1926, reached storage densities of 10^8 bit/cm^2 [2.31]. However, the long access time of microfilm and microfiche information systems make them undesirable for current applications.

---

[2] Note that the term "photorefraction" has two meanings. In the broader interpretation, every light induced change in the complex refraction index $\tilde{n} = n - i\kappa$, is called photorefraction. Therefore, the light induced Kerr effect in liquids or the photochromic effect (light induced change in the absorption constant $\Delta\kappa$) are also photorefraction phenomena. The narrower definition restricts photorefraction to the light induced change of birefringence in electro-optical crystals (Sect. 4.2).

Real time optical recording can be used for optical imaging (i.e., for real time photography) as well as for digital and holographic recordings (Table 2.2). At present, real time optical recording is usually digital on commercial memory discs [2.32]. Computer generated holograms have also been written [2.33] on such discs. However, only amplitude holograms can be produced on commercial optical memory discs with a low diffraction efficieny.

The recording time for direct (i.e., real time) recording must be as short as possible. Another important parameter is the lifetime of the recorded information (optical image, hologram, digital spot). This can be quite different in various situations, from very short (nanoseconds) in fast optical processors, up to long-time (years) in archival memory systems. The light sensitivity of real time recording materials is several orders of magnitude smaller than that of silver halide photoemulsions. These materials require much higher light intensities or longer exposure times than simple photographic films, hence, the necessity for lasers. The light sensitivity of electro-optical materials can be strongly enhanced by recording in an external field. These recording parameters depend on many factors and will be discussed in detail in Chaps. 4 and 5.

Some applications of different recording materials are represented in Table 2.4. At present, most applications are still connected with archival memory systems and optical memory blocks of digital electronic computers. Multilayer Pockels readout optical memory (PROM) cells, developed at the end of the 1970s, allow optical image recognition and processing [2.36,37]. The drawback of PROM cells is their low spatial resolution.

Table 2.4 Applications of optical recording materials in memory systems

Device	Specifications	Memory Systems
High resolution photoplate for holographic recording [2.31, 34]	High resolution (2000–6000 lines/mm); storage density 10^8 bit/cm^2; storage time – decades; high light sensitivity – $5 \cdot 10^{-5}$ J/cm^2	Archival
Optical discs: Compact discs; WORM discs; Reversible discs [2.32,35]	High storage density 10^8 bit/cm^2; recording (erasure) time $0.1 - 1.0$ μs; storage time > 10 years; sensitivity ≤ 10 nJ/bit	Archival; optical memory blocks for digital computers
Pockels readout optical memory cell (PROM) [2.36, 37]	Low resolution; high sensitivity $5 \cdot 10^{-6}$ J/cm^2; rapid digital and holographic recording and erasing ($< 10^{-8}$ s); spatial filtering	Optical processing; image enhancement and restoration
Optical neural networks [2.38, 39]	Interconnection and switching elements (up to 10^{12}); processors	Associative memory systems for following computer generations

In addition to information storage, work is progressing on the development of a pure digital optical computer (DOC) [2.40]. In these machines, all operations are carried out by optical switching – instead of voltages light signals (intensity, polarization) are used. The primary advantage of DOCs isthe extremely short switching time (less than picoseconds), several orders of magnitude faster than for electronic computers. At present, such fast optical processors are not elaborated.

As mentioned in Chap. 1, neural networks are an attractive future develop-ment for associative memory systems similar to the information process in the human brain [2.38]. Such computer systems have an extremely large number of interconnections, (up to 10^{12}), which can be realized only by holographic and opto-electronic devices (HOE, nonlinear processors etc.).

2.4 Holography

In this section we present a brief view of holographic information recording (and readout) and the principles of holographic optical elements which are used in different optoelectronic devices. (For more detailed information see [2.41–46]).

2.4.1 Recording and Readout

Holography is a two step imaging process. The first step in producing a holographic image involves the illumination of the object with a coherent beam of light. The resulting scattered wave emanating from the object is allowed to interfere with a reference wave, whereby the resulting interference pattern is recorded on a photosensitive medium (a photoemulsion requiring subsequent development or a real time recording material). The procedure involved in recording the scattered and reference waves is depicted in Fig. 2.1a: a spherical wave from a pinhole S (the *object or subject light beam*, I_S) and a plane wave (the *reference beam*, I_R). The interference pattern is illustrated in Fig. 2.1b; the concentric interference patterns are similar to that seen on a Fresnel zone plate, shown in Fig. 2.1d. Holograms produced from the interference of a spherical and plane light wave are called *Fresnel holograms* [2.45]. For a more complex object, for example a sculpture, the interference patterns in the hologram are a superposition of the scattered light from different parts of the object, i.e. from every "point" with an order of magnitude of the wavelength of light.

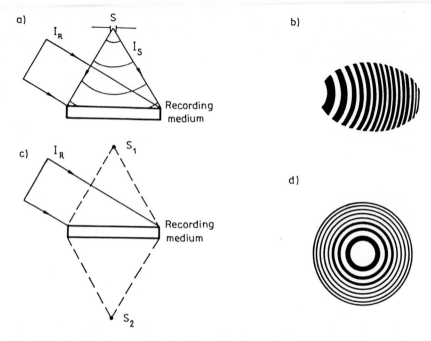

Fig. 2.1a–d. Hologram recording and image reconstruction: (**a**) Fresnel hologram recording by a spherical wave from a pinhole S (subject beam I_S) and a plane wave (reference beam I_R). (**b**) Interference patterns in the hologram. (**c**) By illumination with the reference beam (hologram reconstruction or readout) two images (S_1) and (S_2) are created. (**d**) Interference pattern in a Fresnel zone plate recorded (with I_R perpendicular to the plate) by a similar setup as in (**a**)

The second step of holographic imaging is the reconstruction or readout of the hologram. For this, the hologram is illuminated with the reference beam under the same angle of incidence as that used in the recording step (Fig. 2.1a, c). The scattered light from the hologram produces two images of the object (S_1, S_2) on either side of the hologram. The image S_1 coincides with the object (in our case the point light source S, Fig. 2.1a) and is called the *virtual image*; the second image S_2 on the opposite side is the *real image*.[3]

Holographic images are optical "copies" of the object. Holographic imaging is a wave front reconstruction process, which allows one to obtain a real three dimensional image of volume objects without lenses or objectives. For

---

[3] Two images are produced by a thin hologram (Sect. 2.4.2). These images are equivalent and can be seen with the eye or can be projected onto a screen. The real image is by looking visually inverted and a little bit smeared. Thick or volume holograms have only one three dimensional image. The terms real and virtual images come from D. Gabor who used a coaxial recording scheme in which the virtual image could not be seen.

information processing a multichannel holographic recording and readout is possible which widely increases the field of application .

2.4.2 Hologram Classifications

The type of the hologram depends not only on the photochemical parameters of the recording medium but also on its physical dimensions and the profile of the interference patterns. In the following, we will briefly describe the properties of thin (plane, or two dimensional), thick (volume, or three dimensional), amplitude, phase, transmission and reflection holograms.

As mentioned in Sect. 2.2, any photoinduced effect is the result of a change in the complex index of refraction $\tilde{n}$ or the optical path length $d_{opt} = d \cdot n$. When $\Delta\kappa >> \Delta n$, one speaks of an amplitude recording, since a change in the absorption index is felt most strongly by the amplitude of the transmitted wave. When the phase of the transmitted wave is more strongly modulated than the amplitude, then one speaks of a phase hologram. In this case, $\Delta n >> \Delta\kappa$, or, for the optical path length, $\Delta(d \cdot n) = n\Delta d + d\Delta n \approx n\Delta d$, i.e., when exposure to light produces a change in the profile of the surface of the recording medium (Chap. 4), the optical path length is also changed by variation of Δd, thus, we again have a phase hologram. Such recordings can only be made when the thickness of the medium is large enough to allow a change Δd without burning through the material. Thus, in this case, one speaks of a thickness modulated hologram.

Volume (or 3-D) holograms can also be either phase or amplitude holograms, depending on whether the optical path length or the optical constants (n, κ) of the medium are changed. The reconstruction of a volume hologram, which can be done with a parallel white light beam, produces only a single image: whether it is real or imaginary (virtual) depends on the direction of the incident reference beam. Let us discuss the hologram properties of plane wave interference patterns. The intensity of the subject beam is $I_S \sim A^2_S$ and of the reference beam is $I_R \sim A^2_R$. For plane waves, the interference pattern is thus (Fig. 2.2a)

$$I(x) = A_S^2 + A_R^2 + 2A_S A_R \cos(K_R - K_S)x \,, \tag{2.4}$$

where x is a vector, parallel to the surface of the medium, and

$$K = K_R - K_S \,, \tag{2.5}$$

is the grating vector. K is perpendicular to the planes of the grating and is of length $|K| = 2\pi/\Lambda$.

The period of the grating Λ is related to the angles of incidence θ_S and θ_R (Fig. 2.2a)

$$\Lambda = \frac{\lambda}{\sin\theta_S + \sin\theta_R}, \tag{2.6}$$

where λ is the wavelength of the reference beam outside of the hologram. If $\theta_S = \theta_R = \theta$,

$$\Lambda = \frac{\lambda}{2\sin\theta}. \tag{2.7}$$

Note, when $\theta \to \pi/2$, the period $\Lambda \to \lambda/2$ and when $\theta \to 0$, $\Lambda \to \infty$.

The angles of incidence of the subject and reference beams determine the orientation of the holographic grating, which can be found by a simple vector model $K = K_S + K_R$ (Fig. 2.2). If the object and reference beams are directed at the recording medium from different sides, the interference patterns (and the holographic grating) are parallel to the surface of the medium. The smallest period grating is produced when A_S and A_R are oriented perpendicular to the surface

$$\Lambda_{min} = \frac{\lambda}{2n}, \tag{2.8}$$

where n is the refractive index of the recording medium and λ is the wavelength outside it. Such holograms are called *reflection holograms*. The holographic image is produced on the same side from which the reconstructing beam is incident.

The holograms can be distinguished as thick (3-D) or thin (2-D) by the Q-parameter criterion given by *Klein* [2.44]:

$$Q = \frac{2\pi\lambda d}{n\Lambda^2}, \tag{2.9}$$

where λ is the wavelength of the illuminating reference beam in a vacuum, n is the refractive index of the recording medium and Λ again the period of the holographic grating (Fig. 2.2). Generally, if $Q \geq 10$ the hologram is classified as a volume hologram. Volume holograms are spatial interference filters. This was first shown experimentally and theoretically by *Denisyuk* [2.47] (more details can be found in [2.44]). They are used in "holographic art" to create

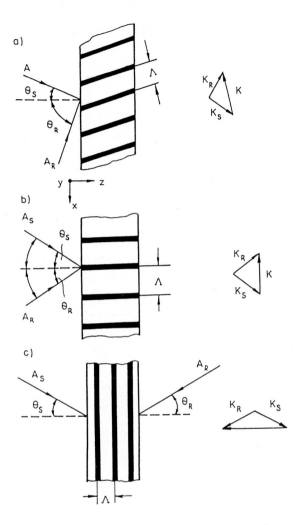

Fig 2.2a–c Interference patterns in transmission (**a**, **b**) and reflection (**c**) holograms: A_S, A_R - subject and reference waves; θ_S, θ_R - angles of incidence for the subject and refernce beams; K_S, K_R, K - wave vectors of the subject and reference beam and the holographic grating

ethereal images of three dimensional objects. *Benton* [2.48] invented the rainbow phase hologram. These are recorded with a special dispersion element so that in viewing, the color changes with viewing angle.

The region $1 \leq Q \leq 10$ represents intermediate holograms, the properties of which are between 3-D and 2-D holograms. However, such quasi-volume holograms can be used in many applications (HOE, phase gratings, etc.). For thin holograms (2-D) $Q \leq 1$ and these holograms can be reconstructed only with coherent laser light.

Holographic Gratings. When the fringes of a holographic grating vary sinusoidally the changes in the absorption index $\kappa(x)$ and the index of refraction $n(x)$ are given by

$$\kappa(x) = \kappa_0 + \kappa_A \cos \boldsymbol{K}\boldsymbol{x}, \tag{2.10}$$

$$n(x) = n_0 + n_A \cos \boldsymbol{K}\boldsymbol{x}. \tag{2.11}$$

Here κ_0 and n_0 are the average values of the absorption and refraction indices after exposure and x is parallel to the surface of the medium (Fig. 2.3a,b). These expressions are valid only for processes that respond linearly to the exposure $(I \cdot t)$ and for sinusoidal gratings.

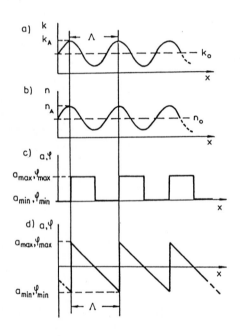

Fig. 2.3a–d. Holographic grating modes. (a, b) Amplitude and phase holograms with sinusoidal modulation. (c) Rectangular (amplitude or phase). (d) Sawtooth-like (amplitude or phase)

In general, however, any hologram can be expressed by the superposition of elementary sinusoidal gratings:

$$q = \sum_{i=1}^{i_{max}} q_{A,i} \cos \boldsymbol{K}_i \boldsymbol{x}, \tag{2.12}$$

where q is either n or k. Figure 2.4 shows realistic patterns for Δn in holographic gratings conditions and a non-sinusoidal hologram after higher exposure [2.49].

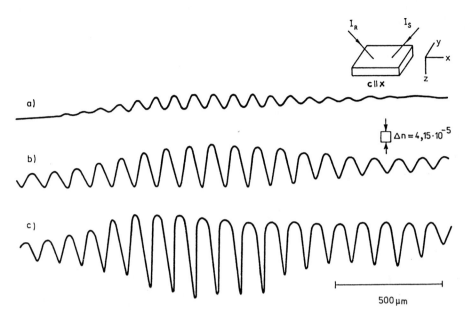

Fig. 2.4a–c. The spatial distribution of the refraction index of a phase hologram in LiNbO$_3$-Fe crystals at different exposures (the recording arrangement is shown in the inset): (**a**) 12 J/cm^2; (**b**) 42 J/cm^2; (**c**) 190 J/cm^2. The phase grating in (**a**) is modulated by the Gaussian beam distribution (therefore the magnitude of Δn in the center is larger than in the periphery). With higher exposures (**b, c**) nonlinear effects are observed

The properties of non-sinusoidal gratings are analyzed in [2.44, 46]. Rectangular and sawtooth gratings (Fig. 2.3c, d) can be produced by photon or electron beam lithography. These gratings are useful in some specialized applications (computer generated holograms, holographic optical elements).

2.4.3 Diffraction Efficiency

One of the main hologram characteristics is the diffraction efficieny η. In the general case, the diffraction efficiency is determined by the relation of the power of the diffracted light beam (P_{diffr}) to the incident power of the beam P_{inc}:

$$\eta = \frac{P_{diffr}}{P_{inc}} . \tag{2.13}$$

The power of the beam is determined by the integral of the light intensity (I) over the surface of the detector (S):

$$P = \iint_S I dS. \tag{2.14}$$

In the case of lasers with Gaussian light beams, the relation between the mean light intensity $(\bar{I})$ and power (P) is described by:

$$\bar{I} \approx \frac{P}{\pi r^2} ,$$

(2.15)

where r is the Gaussian radius of the beam (Sect. 3.2.2). If the light intensity of the subject and reference beams are I_S and I_R, and the diffracted beam intensities are I'_S, I'_R (Fig. 2.5) then the diffraction efficiency is given by:

$$\eta = \frac{I'_{S(R)}}{I_{R(S)}}.$$

(2.16)

Table 2.5. Diffraction efficiency for different holographic transmission gratings [2.44, 49–51]

Dim.	Profile	Equation for diffraction efficiency	η_{max}, %
		Amplitude holograms	
3-D	sinusoidal	$\eta = \exp[(-2\kappa_0 d)/\cos\theta_i]\cdot\text{sh}^2[\kappa_A d/(4\cos\theta_i)]$ $\eta \to \eta_{max}$ at $\kappa_A d/\cos\theta_i = 2\ln 3$ and $\kappa_A = \kappa_0$	3.7
2-D	sinusoidal	$\eta = (t_0^2 m^2/4)\cdot\cos^3\theta$ $\eta \to \eta_{max}$ at $t_0 = 1/2$, $m = 1$ and $\theta \to 0$	6.25
2-D	rectangular	$\eta = \pi^{-2}\cdot\cos^3\theta$ η_{max} at $\theta \to 0$	10.1
		Phase holograms	
3-D	sinusoidal	$\eta = \sin^2(\pi n_A d/\lambda\cos\theta_i)$ $\eta \to \eta_{max}$ at $n_A d/\cos\theta_i = (\lambda/2)(2l+1)$ $l = 0, 1, 2...$	100
2-D	sinusoidal	$\eta = J_1^2(2\pi n_A d/\lambda\cos\theta_i)$ $\eta \to \eta_{max}$ at $(2\pi n_A d/\lambda\cos\theta_i) \to 2$	33.0
2-D	rectangular	$\eta_{max} = (4/\pi^2)\cos^3\theta$ $\varphi_{max} = \pi$, $\varphi_{min} = 0$; η at $\theta = 0$	40.5
2-D	sawtooth	$\eta_{max} = t_0^2\cos^3\theta$ $\varphi_{max} = \pi$, $\varphi_{min} = -\pi$; η_{max} at $t_0 = 1$ and $\theta = 0$	100

η – diffraction efficiency; t_0 amplitude transmittance; m modulation factor ($m = t_A/t_0$); κ_A – amplitude of the modulated index of absorption; κ_0 mean index of absorption of the recording medium after exposure; d - thickness of the medium; θ, θ_i - angles of incidence in a vacuum and in the medium, respectively; J_1 – the first order Bessel function; n_A – refractive index; n_0 – mean refractive index of the recording medium after exposure; λ – the wavelength of light in a vacuum

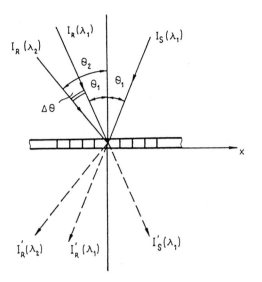

Fig. 2.5 Recording and reconstruction of a holographic grating; $I_S(\lambda_1)$, $I_R(\lambda_1)$ – subject and reference beams; $I'_S(\lambda_1)$, $I'_R(\lambda_1)$, $I'_R(\lambda_2)$ – reconstructed beams; λ_1 is the recording wavelength, λ_2 the reconstruction wavelength. For reconstruction at a different wavelength the angle of incidence must be changed to $\theta_2 = \theta_1 + \Delta\theta$

The diffraction efficiency depends on the wavelength as well as on the thickness of the recording medium. For different elementary holographic gratings the diffraction efficiency is summarized in Table 2.5 (see also Fig. 2.3). The diffraction efficiency depends on whether the hologram is 3-D or 2-D, phase or amplitude, and on the type of grating profile. Note that the brightness of the images made in amplitude holograms does not depend very strongly on the thickness of the medium – the maximum for 3-D holograms is 3.7%, which is smaller than for 2-D, 6.25%, Table 2.5. Amplitude recordings have a much lower diffraction effeciency than phase holograms which, of course, makes them less attractive. The diffraction efficiency can be used to determine the type (amplitude/phase, thick/thin) of hologram being scanned.

In general, however, the real time recording leads to mixed amplitude/phase holograms. For high efficiency hologram recording the absorption of the medium (at the recording wavelength λ_1) must be sufficiently strong. The properties of the recorded hologram depend on the wavelength during reconstruction, λ_2. Upon reconstruction with light of the same wavelength ($\lambda_1 = \lambda_2$), the hologram is usually an amplitude and during readout the holographic properties can be changed by the photoactive light. Therefore, it is more common for the reconstruction light to be of longer wavelength ($\lambda_2 > \lambda_1$). In such a case, the hologram is usually a phase hologram with a higher diffraction efficiency and the readout process would not change the properties of the recorded hologram (i.e., the light of λ_2 is not photoactive). However, for optimal reconstruction efficiency at $\lambda_2 \neq \lambda_1$, the angle of incidence for the reference beam during readout must be changed from θ_1 to $\theta_2 = \theta_1 + \Delta\theta$ (Fig. 2.5).

According to (2.7), θ_2 is given by:

$$\theta_2 = \arcsin\frac{\lambda_2}{2\Lambda} = \arcsin\left[\frac{\lambda_2}{\lambda_1}\sin\theta_1\right] \tag{2.17}$$

(the grating period Λ is determinated by λ_1 and θ_1). For light sensitive As$_2$S$_3$ films, the optimal wavelengths for recording and readout are $\lambda_1 = 514.5$ nm (Ar$^+$ laser) and $\lambda_2 = 632.8$ nm (He-Ne laser). For grating periods of 0.5–1.0 µm and $\lambda_1 \neq \lambda_2$, the angle correction $\Delta\theta/\theta$ is approximately 20%. Under such conditions, in 10 µm thick films, diffraction efficiency of 80% was reached which is close to the theoretical limit (Sect. 4.3). The diffraction efficiency is smaller than the theoretical maximum (100 % for 3-D phase holograms) when the amount of reflected light increases. Thus, materials with high refractive indices (e.g., for As$_2$S$_3$ films $n = 2.49$ with a reflection coefficient of 18%) must be coated with antireflecting layers. This is particularly important for optoelectronic devices [2.32, 51, 52].

Note that η is a function of the amplitude of the induced holographic grating (κ_A, n_A, Table 2.5, Fig. 2.3) which depends on the length of exposure and it´s intensity $I \cdot t$. This is shown in Fig. 2.6 for a phase hologram in a LiNbO$_3$-Fe crystal (η is a periodic function of n_A).

2.4.4 Resolution and Storage Capacity

Holograms are, in general, a superposition of several amplitudes and/or phase gratings which, when read out, reconstruct the volume image, i.e., the wavefront of the scattered light from the object. Therefore, the hologram can be regarded as an optical device, with associated aberrations and diffraction limited resolution (spectral, angular, spatial). A detailed analysis of potential aberrations and the resolution of optical systems is given in [2.19]. Here, we will analyze only some general properties of 3-D holograms which determine their applications in information storage and optical processing.

Spectral Resolution. The spectral resolution of 3-D holograms depends on the indices of refraction, absorption of the medium, thickness, and on the angle of incidence. For a 3-D diffraction grating, the limited spectral region ($\Delta\lambda_{max}$) is determined as [2.44]:

$$\Delta\lambda_{max} = \frac{\Lambda}{d} \cdot \lambda \cdot \cot\theta, \tag{2.18}$$

where θ is the angle of incidence for I_S and I_R ($\theta_S = \theta_R = \theta$, Fig. 2.2). Thus, the available spectral region depends on the ratio: Λ/d. In most applications,

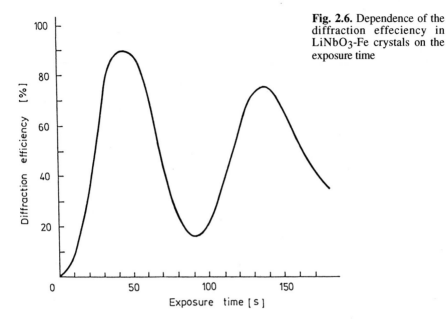

Fig. 2.6. Dependence of the diffraction effeciency in $LiNbO_3$-Fe crystals on the exposure time

$\Delta\lambda_{max}$ is determined by the thickness, d, since the period of the grating is usually $\Lambda \approx \lambda$, which is much smaller than the thickness of the recording medium for 3-D holograms. This can be illustrated in the case of $LiNbO_3$ with $d = 1$ cm; $\theta = 4°$; $\lambda = 500$ nm and $\Lambda = 1$ μm, then $\Delta\lambda_{max} \approx 0.8$ nm [2.52]. Consequently, the reconstruction of the holographic image will require light with the same wavelength. Intermediate holograms in As_2S_3 films ($d = 10$ μm; $\theta = 16°$; $\lambda = 500$ nm) with a grating of $\Lambda = 1$ μm have a larger magnitude of $\Delta\lambda_{max} \approx 180$ nm, thus the wavelength can be changed for the reconstruction of the image. This is an advantage of intermediate holograms for high efficiency phase hologram recording (Sect. 4.3). On the other hand, thick phase holograms ($d \approx 0.1-1.0$ cm) with a small $\Delta\lambda_{max}$ are used as narrowband interference filters [2.34, 44, 45]. However, the most important application of 3-D holograms is pictorial holography [2.44, 47]. For a reasonable good image, the thickness of the recording material (silver halide emulsion, thermoplastics etc.) must be of the order of 20–30 μm.

Angular Resolution. The angular resolution of 3-D holograms determines the number of holograms that can be accomodated in a medium, i.e., a recording of many holograms under different angles of incidence (Fig. 2.7). Such a multistorage method is widely used in holographic memory systems (data storage, coherent filters etc.) [2.43, 44, 53].

Under optimum conditions, it should be possible to separate out each recorded hologram, i.e., the holograms must be recorded at angles the

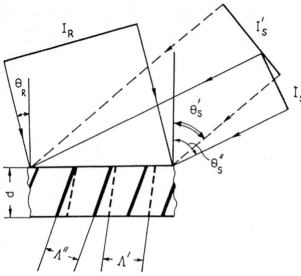

Fig. 2.7. Multihologram recording under different angles of incidence

difference of which is larger than the angular resolution, δ. If the diffraction efficiency is $\eta(\theta)$ at the angle of incidence θ, then $\eta(\theta \pm \delta) \to 0$, that is, holograms recorded at angles $\theta \pm \delta$ will not interfere with one another. According to *Kogelnik,* the angular resolution, δ (in radians), is given by [2.44, 50]:

$$\delta = \frac{\lambda}{2\bar{n}d\sin\theta} \approx \frac{\Lambda}{d} \, , \tag{2.19}$$

where $\bar{n}$ is the average refractive index of the recording medium. The angular selectivity of two amorphous thin films of As_2S_3 with $d_1 = 11$ μm and $d_2 = 5$ μm thickness have been measured by the authors (cf. Fig. 2.8). The estimated halfwidth for the 11 μm film was $\delta \approx 5°$, which is in good agreement with (2.19). The angular resolution of the 5 μm film, with $\Lambda = 0.7$ μm, is larger. However, the 5 μm film is not a 3-D hologram and, therefore, its angular resolution can not be estimated by (2.19).

For a 1 mm thick $LiNbO_3$ sample containing a holographic grating of $\Lambda = 1$ μm, the angular selectivity, calculated from (2.19), is $\delta \approx 0.1°$. This allows one to record up to 900 holograms at different incident angles. *Krätzig* and *Kurz,* under similar conditions, recorded 300 holograms in $LiNbO_3$-Fe crystals [2.53, 54].

Spatial Resolution of a hologram is determined by two general factors: 1. the diffraction of light: 2. the resolution of the recording medium. The first factor – light diffraction, is common for every optical system (lenses, diffraction grating). The spatial resolution of such optical systems is described by the

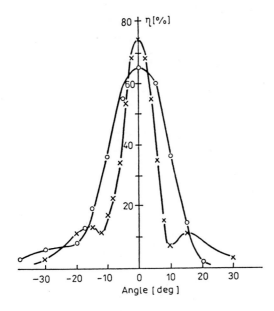

Fig. 2.8. Angular selectivity of AS$_2$S$_3$ films (thickness 11 μm (*crosses*) and 5 μm (*circles*)) for a grating period of $\Lambda = 0.7$ μm (recording wavelength $\lambda_1 = 514.5$ nm; reconstruction wavelength $\lambda_2 = 632.8$ nm). The secondary diffraction maxima on the thicker film are caused by overlap of higher diffraction order

Rayleigh criterion and the Abbe theory [2.19]. According to this theory, the minimal distance which can be resolved on the hologram is equal to :

$$d_{min} = \frac{\lambda}{n \sin \theta} , \tag{2.20}$$

where λ is the wavelength outside, n is the refraction (i.e., the half of the angle between the border rays from the pinhole to the hologram, Fig. 2.1). As $\theta \to 90°$, $d_{min} \approx \lambda / n$, which is equal to the wavelength in the recording medium. Thus, if we record an elemental holographic grating with a period Λ, the diffraction limitations require that:

$$\Lambda \geq d_{min} \approx \frac{\lambda}{n}. \tag{2.21}$$

However, for the recording of a holographic grating with $\Lambda \geq d_{min}$, it is necessary that the structural resolution of the recording medium, $d_{rec,}$ is equal or better than d_{min}.

The spatial resolution of the recording medium depends on the atomic structure of the material (the evenness in the distribution of the photoactive sites), as well as on the macroscopic inhomogeneities in the bulk and on the degree of surface roughness and flatness. Typical spatial resolutions in commercially available photographic emulsions (Agfa-Gevaert, Eastman-Kodak) are 5000 lines/mm, corresponding to $d_{rec} = 0.2$ µm (i.e. $d_{rec} < d_{min}$). This is the physical limit for silver halide photoemulsion caused by grain structure of elementary light sensitive areas after development [2.34].

The spatial resolution of the recording medium is an important factor for applications in photo- and electron beam lithography. The best real time recording materials have a higher resolution than silver halide photoemulsions, e.g., As_2S_3 films have a value of $d_{rec} \leq 0.1$ µm (Sect. 4.3).

Storage Capacity. The storage density of a hologram depends not only on the properties of the recording medium but also on the resolution of the other optical components in the system, e.g., the laser beam width, as well as lens and mirror qualitiy. In order to be able to produce a hologram grating with a period of Λ_{min}, all spatial fluctuations of the light beams must be smaller than Λ_{min}.

If we realistically assign the best possible resolution of a recording apparatus as $\Lambda_{min} \approx \lambda$, taking the optical system into account , we get, from *van Heerden* [2.5], the maximum storage density for a 2-D hologram as:

$$N_{2D}^{max} = \frac{1}{\lambda^2} \tag{2.22}$$

and for a 3-D hologram

$$N_{3D}^{max} = \frac{1}{\lambda^3} . \tag{2.23}$$

Thus, for an illuminating wavelength of 500 nm, this yields $4 \cdot 10^8$ bit/cm^2 and $8 \cdot 10^{12}$ bit/cm^3. The same diffraction limitations exist also for digital recordings. For commercial plane optical memory discs, typically $\sim 10^8$ bit/cm^2, for a total capacity of $6 \cdot 10^9$ bit on a 130 mm diameter disc [2.35].

2.4.5 Holographic Optical Elements

Optical elements such as lenses, beam splitters, diffraction gratings and filters can be produced by holographic imaging. These holographic optical elements (HOE) have the advantage of being cheap (due to their simple design, small size, low weight) and are easily reproducible by pressing polymer materials. They are wavelength selective and have a high diffraction efficiency, corresponding to large apertures of conventional lenses.

HOEs are most important in optoelectronics, in particular, for optical memory discs and diode lasers. The latter account for nearly 90% of the world's production of lasers [2.35, 55]. Diode lasers have a very large angle of divergence, which can be compensated by the installment of HOE focusing elements.

The simplest HOE is a Fresnel zone plate which can act as a focusing lense (see the Fresnel hologram recording in Fig. 2.1). If the HOE is illuminated by a reference beam of I_R at the same angle of incidence as the recording, it can act as a classical optical element.

In general, off-axis HOEs (Fig. 2.9) are more useful than the on-axis variety. Reconstructing off-axis, not only reduces the reflection losses but also avoids interference from higher diffraction orders.

The theoretical diffraction efficiency of a 3-D transmission phase HOE is 100% (Table 2.5). The actual η is less than this, however, and can be

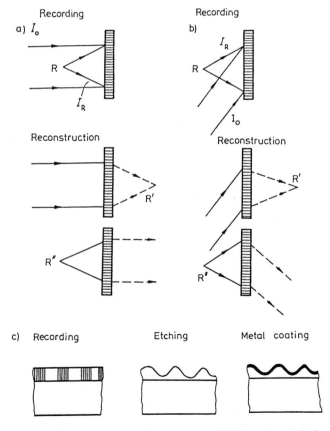

Fig. 2.9a–c. Holographic optical elements (HOE): **(a)** transmission axial; **(b)** transmission off-axis (nonaxial); **(c)** three steps in producing a surface relief reflection HOE

improved only when the medium is coated with an antireflection layer. Reflection losses can be decreased by using coatings with lower refractive indices (for glasses $n = 1.5$ and for normal incidence the reflection coefficient R is 4%; for materials with $n = 2.5$, the reflection coefficient is $R = 18\%$).

The materials used in HOE are subject to the same imperfections as conventional optical elements (lenses, objectives). Mimimum aberrations can be achieved only when the image is reconstructed at the same wavelength and the same angle of incidence as that at which it was recorded. This is not usually possible in real time recording materials. In order to obtain high diffraction efficiencies, the readout must be done at another wavelength ($\lambda_1 \neq \lambda_2$) which leads to a drastic increase in aberrations. Conventional silver halide materials also have some limitations for HOE production, caused by a change in the interference pattern structure during chemical processing (development). Therefore, special recording materials are necessary for HOE production.

One of the best materials for HOE production are photoresists. In these materials, the primary holographic recording is followed by etching to produce a surface profile which is additionally covered with a metal layer (Fig. 2.9c). Significant progress is obtained in computer generated holograms using laser beams, X-rays or electron beams [2.56–60].

2.5 Light Sensitivity of Recording Materials

The light sensitivity determines the exposure required for optical imaging. During more than one hundred years (from the time of the invention of photography up to the invention of lasers, Table 2.1) only amplitude recording ($\Delta\kappa > \Delta n$) in conventional photomaterials (silver halide or organic) was used for optical imaging. As has been previously mentioned, the light sensitivity of conventional photomaterials is much higher than for real time recording materials (this is caused by the chemical development procedure introducing a strong amplification of the primary photoinduced effect). Light sensitivity is determined by the absorbed energy which is necessary for a photophysical or photochemical reaction in the recording medium. It is defined by various criteria.

In conventional photographic films the light sensitivity is determined by the criterion of *Hurter and Driffield* [2.61] which uses the dependence of the optical density D on the length of exposure and its intensity $I \cdot t$ (Fig. 2.10). At low exposures, no measurable change in the optical density occurs (Fig. 2.10, region 1).

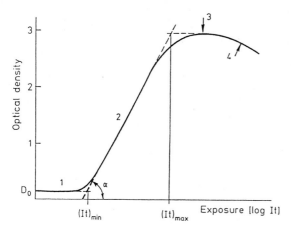

Fig. 2.10. Typical optical density curve for an optical recording material: (*1*) low exposure region (D_0 is the fog), (*2*) the linear exposure region; (*3*) saturation; (*4*) solarization; $\tan\alpha$ is the contrast coefficient

The quantity D_0 is the "fog" of the exposed film (or a real time recording material), which determinates the smallest measurable light induced optical density change ($\Delta D > D_0$) and the noise[4] level of the recording material. At higher exposures, the relationship to $I \cdot t$ is linear; the slope of the line is used in the calculation of the light sensitivity (Fig. 2.10, region 2). In photomaterials, D reaches a maximum at $(I \cdot t)_{max}$, and begins to decrease at higher exposures (Fig. 2.10, regions 3 and 4). This is due to saturation of the primary light sensitivity centers in the material. Overexposure deteriorates the optical image. The change in the optical density D (or the light induced changes in optical constant $\Delta\kappa$ or Δn) is a general effect which occurs in many real time optical recording materials (Chap. 4).

Along with the light sensitivity, we can define several other useful parameters: the contrast coefficient γ; the minimum detectable optical density above the fog level $D_0 + \Delta D$; the length of the linear response region from the exposure $(It)_{min}$ to $(It)_{max}$. The coefficient of contrast is given by:

$$\gamma = \tan\alpha = \frac{\Delta D}{\Delta\log(I \cdot t)} \tag{2.24}$$

and characterizes the rate of the photoinduced effect, i.e. the slope of the curve. The linear range $\Delta(I \cdot t) = (I \cdot t)_{max} - (I \cdot t)_{min}$ determines the linear response region of the recording medium. This parameter is particularly important in knowing when to make multiple holographic images so as to optimize the exposure for each hologram and to avoid overexposure.

---

[4] The noise results from inhomogeneities of the materieal, as well as from the light source (e.g., speckle noise) [2.34, 61].

The sensitivity of the material is determined by the exposure $(I \cdot t)_{\Delta D}$ at which the optical density is greater than D_0 by ΔD,

$$S_{\Delta D} = \frac{1}{(I \cdot t)_{D_0 + \Delta D}} \,. \tag{2.25}$$

The *International Optical Society* has set the convention: $\Delta D = 0.1$, which is used in ASA and DIN specifications. S is measured in cm²/J. Often, the inverse value S^{-1} (J/cm²) is used. In digital recordings on optical memory discs the light sensitivity, S_{dig}^{-1}, is usually found from the absorbed energy per recorded bit [2.62, 63]. In real time optical recording materials, S_{dig}^{-1} is of the order of 0.1–1.0 nJ/bit which corresponds to the inverse light sensitivity value of $S^{-1} \approx 0.01$–0.1 J/cm² (for recorded spots of 1 μm size, Sect. 5.2).

In holographic recording, rather than using the optical density as in photography, the diffraction efficiency, discussed above, is used to determine the light sensitivity. This is a general parameter which can be applied to all optical media (see also [2.34] p. 123). This is illustrated in Fig 2.11 for one of the best silver halide photoplates Kodak G49-F (emulsion thickness 17 μm) used as a holographic medium.

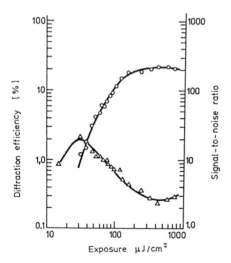

Fig. 2.11. Dependence of diffraction efficiency (*circles*) and signal-to-noise-ratio (*triangles*) on exposure for a diffuse test object of recorded and bleached Kodak 649-F plate (thickness of the photo-emulsion $d = 17$ μm; $\lambda = 632.8$ nm)

For phase recording in conventional photoemulsions, a special chemical treatment, bleaching of the recorded amplitude hologram, is necessary [2.34]. The maximum diffraction efficiency ($\eta_{max} \approx 20\%$) is reached in the saturation region where the signal-to-noise ratio (SNR) is 3:1 and the light sensitivity is $S^{-1} \approx 4 \cdot 10^{-4}$ J/cm². Along the linear part of $\eta = f(I \cdot t)$, the diffraction efficiency in the optimal SNR region (20:1) is $\eta \approx 2\%$ and the light sensitivity $S^{-1} \approx 4 \cdot 10^{-5}$ J/cm². At present, this is the highest light sensitivity achieved with high resolution optical imaging in silver halide photomaterials. The light sensitivity for real time optical recordings, as mentioned above, is several orders of magnitude smaller (Table 2.6). Usually, the light sensitivity of materials for holographic recording ($S_{1\%}^{-1}$) is determined per 1% diffraction efficiency, and $S_{1\%}^{-1}$ is equal to:

$$S_{1\%}^{-1} = \frac{I \cdot t}{\eta}, \tag{2.26}$$

where $I \cdot t$ is the exposure and η is the diffraction efficiency (in %).

We have proposed some general criterion for the estimation of light sensitivity [2.64]. This criterion is based on the response of the material to holographic recording and can be applied to every recording medium and any recording mechanism. It is common to record an elemental holographic grating in order to estimate the light sensitivity. The minimal grating period which can be recorded also gives information about the spatial resolution of the medium. Using light with different wavelength (λ) the dependence of $S(\lambda)$ can be easily estimated.

The proposed method is illustrated in Fig. 2.12. Two different values for the light sensitivity are defined as that with the highest recording rate (S_A^{-1} at $(It)_{A, max}$) and that with the largest diffraction efficiency (S_B^{-1} at $(It)_{B, max}$). In Fig. 2.6 the first maximum of $\eta = f(It)$ is the largest and includes both sensitivities S_A^{-1} an S_B^{-1}, i.e. $S_A^{-1} = S_B^{-1}$. The light sensitivities S_A^{-1} and S_B^{-1} can also be related to the 1% diffraction efficiency level, which leads to

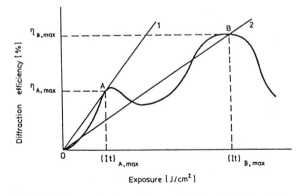

Fig. 2.12. Determination of the sensitivity of light in a holographic recording medium. Point **A** marks the value when the sensitivity should be measured for the highest recording rate (i.e. the largest tangent); point **B** is at the highest diffraction efficiency

$$S_{A1\%}^{-1} = \frac{(I \cdot t)_A}{\eta_{A,\,max}},$$ (2.27)

and

$$S_{B1\%}^{-1} = \frac{(I \cdot t)_B}{\eta_{B,\,max}}.$$ (2.28)

Table 2.6 shows the result of the criteria for several media. We will use this method in the estimation of the sensitivity of light for real time optical recording materials (Chap. 4).

Table 2.6 Light sensitivity of different holographic recording materials

Recording media	$\eta_{A,\,max}$ %	$S_{B\,1\%}^{-1}$, J/cm^2	$\eta_{B,\,max}$ %	$S_{B\,1\%}^{-1}$, J/cm^2	Recording conditions
Kodak 649-F silver halide plate $d=17\,\mu m$	20	$2 \cdot 10^{-5}$	–	–	$\lambda_1=\lambda_2=632.8$nm phase hologram (bleached) $T=290$ K
As$_2$S$_3$ film $d=11\,\mu m$	71	0.3	78	0.4	$\lambda_1=514.5$ nm $\lambda_2=632.8$ nm $\Lambda=0.7\,\mu m$; $T=290$ K
KBr-F $N_F\approx10^{17}$ cm^{-3}; addit. colored; $d=1.5$ mm	3.4	14	6.4	75	$\lambda_1=\lambda_2=514.5$ nm $T=450$ K $\Lambda=1.4\,\mu m$
As$_{0.91}$Se$_{0.09}$ film; $d=5\,\mu m$	4.7	4.0	4.7	4.0	$\lambda_1=\lambda_2=632.8$ nm $T=290$ K $\Lambda=1.2\,\mu m$
As$_{0.37}$S$_{0.63}$ film; $d=5\,\mu m$	21	0.5	30	0.67	$\lambda_1=488.0$ nm $\lambda_2=632.8$ nm $T=290$ K
LiNbO$_3$-Fe crystal; $d=0.2$ mm	–	–	85	0.1	$\lambda_1=488.0$ nm $\lambda_2=514.5$ nm $\Lambda=2\,\mu m$ $T=290$ K

T is the temperature of the sample; λ_1, λ_2 are the recording and reconstruction wavelengths; Λ is the period of the holographic gratings; d is the thickness of the medium

3. Lasers, Nonlinear Optics and Photoelectric Detectors

In this chapter we review the basic principles of optoelectronic processes and devices for digital and holographic information recording and readout. Light absorption, some specifics of laser systems (including laser diodes), selected aspects of nonlinear optics (nonlinear absorption, second harmonic generation), light modulation and photoelectric detectors are discussed.

3.1 Absorption and Emission of Light

Light absorption is a general characteristic of all optical materials. No real medium is perfectly transparent for any region of the spectrum and all media show strong absorption in some definite spectral regions. Dielectric materials have two sharp separated absorption bands: (1) the fundamental absorption; (2) the infrared absorption (Fig. 3.1). The fundamental absorption band is due to electronic transitions in a solid (band to band transitions, exciton absorption). Infrared absorption is caused by excitation of lattice vibrations. The main characteristic of the absorption spectrum of dielectrics and semiconductor materials is the energy gap E_g, which can rigorously be described only by quantum mechanics [3.1, 2].

The energy gap is determined by the minimum photon energy ($h v_g = E_g$) needed to create an electron-hole pair in the dielectric or semiconductor material, i.e., the smallest photon energy required for a transition from the valence to the conduction band. In the spectral region $h v \geq E_g$ (i.e., at wavelengths shorter than the energy gap, $\lambda < \lambda_g$) the absorption coefficient is high. In a regular lattice below E_g there are no energy levels. Thus, E_g also determines the optical transmission (i.e., absorption) region of the material, $E_{vib} \leq h v \leq E_g$, where E_{vib} is the vibration limiting energy (for KCl $E_{vib} \approx$ 0.05 eV and the optical transmission extends from 160 nm to 20 µm, Fig. 3.1). The estimation of the energy gap is difficult because of overlapping of the band-to-band and exciton absorption. Often, E_g is estimated by measurements

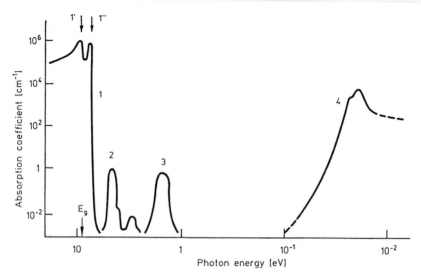

Fig. 3.1. The absorption spectra of KCI crystal: (*1*) fundamental absorption (E_g is the band gap and 1', 1" the exciton bands); (*2*) Tl^+ impurity center absorption; (*3*) F-center absorption; (*4*) infrared vibration spectrum ($E_{vib} = \hbar\omega_{vib}$)

of the onset of photoconductivity or photoemission spectroscopy (the intrinsic photoelectric effect occurs only at $h\nu \geq E_g$) [3.3].

There is no principal difference between the absorption spectra of dielectric and semiconductor materials. However, for semiconductor materials the energy gap is smaller than for dielectrics. Typical dielectric materials (e.g., quartz, oxide glasses and crystalline Al_2O_3, MgO, CaO) are transparent in the UV and visible spectral regions (i.e., $E_g \geq 3$ eV). Semiconductor materials are usually absorbant in the visible spectral region whereas they are transparent in the near infrared region. In Table 3.1 the optical characteristics of some dielectric and semiconductor materials are presented. Below, a quantitative description of optical absorption and emission is given.

The absorption of light results from the interaction of photons with atoms (or molecules) and is characterized by an absorption coefficient κ_a (depending on the light frequency, intensity and polarization), determinated by the electronic properties of the absorber (atom, molecule, solid). The intensity of a light beam I, is proportional to the flux of photons N_ν, where N_ν is the number of photons with energy $h\nu$, per unit area and unit time. Figure 3.2 illustrates the absorption of light by a slab of width Δd. The fraction of the absorbed light intensity ΔI (or number of photons ΔN_ν) equals:

Table 3.1. Absorption coefficients at the indicated wavelengths and transmission regions for some dielectric and semiconductor materials

Material	Energy gap, E_g eV(nm)	Wave-length nm	Absorption coefficient, k_a cm^{-1}	Penetration depth, d_0 μm	Transmission region μm
NaCl (cub.)	8.8(140)	160	10^6	0.01	0.25–16
SiO$_2$ (amorph.)	8.0(155)	190	1.7	$6 \cdot 10^3$	0.2–5
Polymethylmeth-acrylat (PMMA)	–	320	1.2	$8 \cdot 10^3$	0.32–5
GaAs (cub.)	1.35(919)	830	10^4	1	1–15
As$_2$S$_3$ (amorph.)	2.37(524)	477	10^4	1	0.6–13
As$_2$Se$_3$ (amorph.)	1.8(689)	653	10^4	1	0.8–20
α-S (rhomb.)	2.0(620)	354	10^4	1	0.5–30
Se (amorph.)	~1.6(775)	620	$2 \cdot 10^4$	0.5	0.5–30
Se (hex.)	1.79(693)	620	$6 \cdot 10^4$	0.16	1–20
Te (hex.)	0.33(3758)	1000	$1.8 \cdot 10^5$	0.05	4–30
α-As (hex.)	~1.2(1000)	885	10^4	1	1–23
Si (cub.)	1.11(1120)	950	10^2	100	1.2–15
Ge (cub.)	0.67(1851)	1600	10^2	100	3–23

$$\frac{\Delta I}{I} = \frac{\Delta N_v}{N_v} = \frac{\Delta A}{A} \; , \tag{3.1}$$

where ΔA is the total area of the absorbers in the slab of area A. If the volume of the slab is $A\Delta d$ and if the number of absorbers per cm^3 is N_0, then, the magnitude of ΔA is:

$$\Delta A = N_0 \sigma_a A \Delta d \tag{3.2}$$

and

$$\frac{\Delta I}{I} = \frac{\Delta N_v}{N_v} = \sigma_a N_0 \Delta d , \tag{3.3}$$

where σ_a is the absorption cross section. The value of σ_a characterizes the efficiency of the absorbing centers. For strongly absorbing molecules the cross section is approximately equal to the geometrical cross section of a molecule. For weakly absorbing molecules σ_a (in cm^2 or m^2) can be even smaller than 1% of the geometrical cross section.

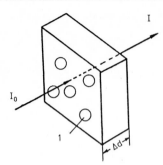

Fig. 3.2. Light absorption in a slab with absorbing particles (*1*) (e.g., molecules, defect centers). I_0 is the incident and I the transmitted light intensity; Δd is the thickness of the slab

To obtain the macroscopic light absorption, we integrate (3.3) to get the Lambert law[1] of classical optics for the transmitted light intensity:

$$I = I_0 \exp(-\sigma_a N_0 d) = I_0 \exp(-\kappa_a d), \tag{3.4}$$

where $\sigma_a N_0 = \kappa_a$ is the linear absorption coefficient in cm^{-1}.[2] The absorption cross section, σ_a, and absorption coefficient, κ_a, depend on the wavelength. This dependence $\kappa_a(\lambda)$, is just the absorption spectrum (Fig. 3.1). As mentioned before, for dielectrics and semiconductors the spectral transmission region begins at $h\nu \leq E_g$ and ends at $h\nu \approx E_{vibr} = h\omega_{vib}$, where ω_{vib} is the limiting phonon frequency of the material (Fig. 3.1 and Table 3.1).

The absorption coefficient determines also the penetration depth of the absorbing light by $\kappa_a d_0 = 1$, i.e., the thickness at which the light intensity decreases inversely with the exponential (3.4). In the fundamental absorption region of κ_a is the order of 10^5 cm^{-1} and $d_0 \approx 0.1$ µm, i.e., smaller than the wavelength of the absorbing light (Fig. 3.1). Therefore, for excitation in the fundamental absorption region only a thin layer of the material near the surface is sensitive to photostimulation (optical recording, luminescence, photo-conductivity, etc.). Thus, for optical recording, usually wavelengths longer than λ_g are used in the tail of the fundamental absorption, color centers or defect states (Fig. 3.1). In contrast, for many nonlinear optical processes, e.g., second harmonic generation, materials with a high transmittance are necessary (Sect. 3.3).

As mentioned before, a correct description of light absorption and emission is possible only by quantum mechanics. This will be done here for a simple three level model (Fig. 3.3). The probability of excitation of the system (an

---

[1] The absorption law was first observed by Pierre Bougier (1698–1758) in 1729 and was further analyzed in detal by Johann Heinrich Lambert. In 1852 Beer showed that in solution the absorption coefficient is proportional to the concentration of the absorbing molecules.

[2] In spectroscopy, the Bougier-Lambert law is written $I = I_0 10^{-\kappa'_a d}$, where $\kappa'_a = 0.4343 \kappa_a$.

atom, molecule, or other quantum mechanical entity) from its lower state i, with energy E_i, to an upper state k, with energy E_k, by a photon with energy $h\nu_{ik} = E_k - E_i$, is proportional to the Einstein coefficient of absorption B_{ik} (for more details see [3.4] Chap. 18). The number of transitions per unit time is:

$$\frac{dN_a}{dt} = N_i \rho_v B_{ik} \, , \tag{3.5}$$

where N_i is the number of absorbers in state i and ρ_v is the the photon energy density per frequency (J·s/cm³).

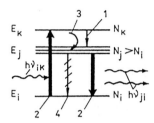

Fig. 3.3. Stimulated and spontaneous transitions in a quantum system: $h\nu_{ik}$ is the energy of the initial incident photon exciting the system (arrow 2) from state i (energy E_i, number of particles N_i) to state k. State k non-radiatively decays to state(s) j via process 3 and relaxes back to the ground state with the emission of a photon with energy $h\nu_{ji}$ (2' is a stimulated and 4 is a spontaneous emission from the metastabile state j). State k can also radiatively relax (processes 1 and 1') by the stimulated or spontaneous emission of a photon. In the case in which N_j or N_k are greater than N_i, there is a population inversion.

Once the quantum system has been excited to state k, it can relax by a number of mechanisms. For example, the energy can be channeled into other modes, such as atomic vibration or rotation. This non-radiative transition is shown by the curved arrow 3 in Fig. 3.3 (for the transition $E_k \rightarrow E_j$). Further transitions back to the ground state may (arrow 2') or may not (arrow 4) be allowed by the rules of quantum mechanics, which are beyond the scope of this book [2.5]. State k can also radiatively relax, that is, by re-emitting either a photon to state j (arrow 1') or back to the ground state i (arrow 1). The system can be induced to relax by the impact of another photon (stimulated emission from $E_k \rightarrow E_i$ by $h\nu_{ki} = E_k - E_i$) or it can do this spontaneously. The emission rates for radiative processes are analogous to the absorption rate:

$$\frac{dN_e^{sp}}{dt} = N_k A_{kj(i)} \, ; \tag{3.6}$$

$$\frac{dN_e^{st}}{dt} = N_k B_{kj(i)} \rho_v \, , \tag{3.7}$$

where j or i indicate the final state of the system for spontaneous and simulated processes. Obviously,

$$B_{ki} = B_{ik} \qquad (3.8)$$

must be true. The relationship between the Einstein coefficients for stimulated B_{ki} and spontaneous A_{ki} processes for the transition is equal to:

$$A_{ki} = 8\pi h v_{ki}^3 B_{ki} / c^3 . \qquad (3.9)$$

The general expression for the total rate of change in the number of particles in the upper state, k, is:

$$\frac{dN_k}{dt} = \frac{dN_a}{dt} - \frac{dN_e^{st}}{dt} - \frac{dN_e^{sp}}{dt} = N_i \rho_v B_{ik} - N_k \rho_v \sum_j B_{kj} - N_k \sum_j A_{kj} , \qquad (3.10)$$

where the last two terms account for transitions to all possible final states j. When $dN_k / dt = 0$, the system is saturated.

3.2 Lasers

3.2.1 Population Inversion

The transition rate depends on the populations in the energy levels (N_i, N_j and N_k) and the photon energy density, ρ_v, which is proportional to the exciting light intensity or spontaneous emitted photons. In Fig. 3.3 a three level quantum system was analysed by the authors by excitation from energy level E_i to E_k (induced absorption transitions described by B_{ik}). The stimulated emission from level E_j can be observed only when the population of level E_j is larger than for level E_i (i.e., if $N_j > N_i$). This is possible only in quantum systems with more than two levels. The energy level population is described by the Boltzmann distribution:

$$N_j = N_0 \exp\left(-\frac{E_j}{k_B T}\right), \qquad (3.11)$$

where N_0 is the number of particles (in our case, electrons on atomic or solid state energy levels) in the ground state ($N_0 = N_i$ in Fig. 3.3), T is the temperature, and $k_B = 1.3805 \cdot 10^{-23}$ J·K^{-1} is the Boltzmann constant. According to (3.11), at thermal equilibrium $N_j < N_0$. In order to incite stimulated emission (i.e., lasing) the rate of emission must be larger than the rate of absorption , i.e., one requires $N_j > N_i$. When this happens, one speaks of a "population inversion" or of a Boltzmann system with a "negative temperature", $T^* = -T$. Such population inversions can be achieved only in quantum systems with more than two levels, when the lifetime of the upper state (in our case E_j) $\tau_j = 1/A_{ji}$ is very long. Such metastable[3] levels are widely used in lasers. Population inversion is reached by excitation from E_i to E_k with a subsequent relaxation from E_k to E_j. For strong inversion, high photon fluxes, ρ_v, or high pumping light intensities are necessary.

Population inversion results in a negative light absorption coefficient $\kappa_{aij} = \sigma_{ij}(N_i - N_j)$ since $N_i - N_j < 0$, and it becomes more negative with increasing N_j . The negative absorption can be used for the quantitative measurement of the population inversion, first observed by *Wavilow* and *Lewschin* in uranyl glasses [3.6].

3.2.2 Principle of Operation

Laser radiation is simply the sustained stimulated emission from excited metastable quantum systems. Lasers consist of several components: (1) an active medium; (2) a means of excitation (e.g., light, gaseous discharge, injection of charge carriers); (3) a resonator and (4) a mode selection system. Different classification systems for lasers are used [3.5], the most common is according to the active media, e.g., gaseous, solid state, dye solutions. A special class of solid state lasers is that of semiconductor or diode lasers.

We will discuss the above four components using the example of a solid state ruby laser (Al_2O_3-Cr^{+3}). The active part of the ruby laser, the photon emitter, is the chromium cation impurity. Cr^{+3} has three groups of electronic states that participate in the lasing: the ground state ($^4A_2$), two excited ($^4F_1$ and $^4F_2$) and two metastable (2 $\overline{A}$ and $\overline{E}$; Fig. 3.4a). The excitation of the Cr^{+3} ions occurs by light absorption in the $^4F_1$ and $^4F_2$ absorption bands (with maxima at $\lambda_{a1} = 410$ and $\lambda_{a2} = 550$ nm). After relaxation to the metastable level $\overline{E}$ (with a lifetime of approximately 3 ms), a population inversion $N_{\overline{E}} > N_{^4A_2}$ is achieved. When the Cr^{+3} ions in the $\overline{E}$ state spontaneously emit, the released photon can impinge on another excited Cr^{+3} ion center to emit a second photon in a stimulated transition. Now the two photons can stimulate induced emission transitions from other Cr^{+3} ions. The result of such a process is a cascade of photons. In a ruby laser, the energy of the photons is

---

[3] The lifetimes of metastable states are $\tau > 10^{-6}$ s which is much longer than for states with allowed transitions (10^{-9}–10^{-8} s).

$hv = 1.8$ eV ($\lambda_{R_1} = 694.3$ nm). Emission from the slightly higher lying $2\overline{A}$ metastable level $\lambda_{R_2} = 692.8$ nm, is not seen except when the λ_{R_1} can be suppressed by using a special spectral selection in the resonator. Pulses and continuous wave emission can be obtained by pulsed or continuous pumping of the system.

To increase the efficiency of the emission cascade formation, the lasing medium is placed into an *optical resonator* which, in the simplest case, consists of two reflectors facing one another (Fig. 3.4b). Only the stimulated emission parallel to the central axis is amplified (the rest is lost out of the sides of the resonator cavity). Part of the amplified light leaks out of the partially reflecting mirror M_2, and is the laser output, I_{out}.

Furthermore, only certain wavelengths of the light moving parallel to the resonator axis will be amplified, namely, those that can form standing waves, i.e., $m\lambda_m = 2\,l$, or in frequency modes $v_m = m \cdot c\,/(2l)$, where m is an integer, c the speed of light and l the length of the resonant cavity. This is illustrated in Fig. 3.5. Part a of the figure shows a typical luminescent spectrum (for example, the emission of excited Cr^{+3} ion). Figure 3.5b shows the possible resonator frequency modes (v_m) and Fig. 3.5c shows, thus, the resulting longitudinal mode emission, i.e., the output of the laser.

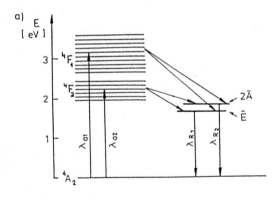

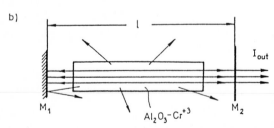

Fig. 3.4. (a) Three level Al_2O_3-Cr^{+3} laser: $^4A_2$ Cr^{+3} ground state; $^4F_2$, $^4F_1$ excited levels; $\overline{E}$, $2\overline{A}$ metastable states (with a lifetime of approximately 3 ms); excitation is achieved with $\lambda_{a1} =$ 410 nm and $\lambda_{a2} = 550$ nm light (the maximum of the bands) and the lasing emission is at 694.3 nm from level $\overline{E}$ and 692.8 nm from level $2\overline{A}$ (the latter is possible only in a resonator which suppresses the main 694.3 nm emission). **(b)** A resonator with two parallel mirrors: M_1 – reflecting mirror; M_2 – output mirror with partial reflection; l – the resonator length. Only rays emitted perpendicularly to the mirrors (parallel to the axis) are amplified and emitted

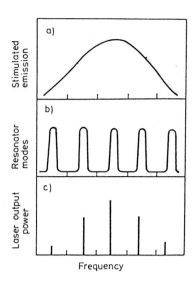

Fig. 3.5a–c. Spectral characteristics of laser emission: (**a**) Stimulated emission from a lasing medium. (**b**) Modes of the optical resonator with frequencies of $v_m = m \cdot c/(2l)$. (**c**) Resulting output modes

The profile of the electromagnetic field of these spectral modes can have various shapes, i.e., transverse modes. A more detailed treatment will not be undertaken here, rather, a few will be illustrated, cf. Fig. 3.6, which shows the light intensity distributions through the cross section of the output beam. The TEM_{00} mode shown in the upper left corner, is sometimes called the axial or fundamental mode and has a Gaussian intensity distribution:

$$I(r) = I_0 \exp{(-2r^2/w_0^2)}, \tag{3.12}$$

where I_0 is the intensity in the center ($r = 0$), r is the distance from the center of the beam and w_0 is the Gaussian radius (when $r = w_0$, the intensity is $1/e^2$ times the intensity in the center). The beam propagates outside the resonator retaining the general Gaussian beam distribution. However, at a distance z from the output mirror a correction to the distribution function must be made, i.e., in (3.12) the Gaussian radius must be replaced by $w(z)$ [3.5]

$$w(z) = w_0 [1 + (\lambda z/\pi w_0^2)^2]^{1/2}. \tag{3.13}$$

At large z (i.e., if the second term in the bracket is much larger than one), the Gauss angle of divergence of the beam, θ_{div}, is equal to:

$$\theta_{div} = \frac{\lambda}{\pi w_0}, \tag{3.14}$$

which is the far-field divergence of a Gaussian light beam.

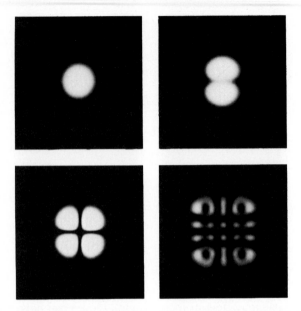

Fig. 3.6. Transverse modes in an optical resonator: upper left, 00 mode; upper right, 01 mode; lower left, 11 mode; lower right, coherent superposition of two or more transverse modes [3.5]

An important laser parameter for applications in holography is the coherence length (l_{coh}) which is determined by the time (τ_{coh}) of coherent light emission. The coherence length l_{coh} also determines the maximum optical path length difference in holography or any other optical interference system (i.e., the optical path length difference Δl_{opt} can not be larger than l_{coh}). The coherence length depends on the linewidth of the laser emission, and is given by:

$$l_{coh} = c \cdot \tau_{coh} = \frac{c}{\delta v} = \frac{\lambda^2}{\delta \lambda} \, , \qquad (3.15)$$

where c is the velocity of light and δv (or $\delta \lambda$) is the linewidth (note that the linewidth $\delta v = 2\delta v_{1/2}$, where $\delta v_{1/2}$ is the halfwidth of the emission line). Therefore, for high-coherence laser emission (i.e., for large l_{coh}) mode selection of the laser emission must be implemented (i.e., selection of one of the output modes in Fig. 3.5c). Such mode selection can obtained by using internal spectral analyzers (prisms, diffraction gratings) or high resolution laser spectroscopy such as Fabry-Perot interferometer. Selection of transverse modes requires careful adjustment of the length of the resonator, mirror curvature and the inclusion of limiting apertures within the cavity.

For different laser applications, various mode selection methods are used. *Single frequency lasers* (one spectral and one transversal mode) have the highest coherence length. For single frequency He-Ne lasers, the coherence length is more than 100 m (Table 3.2).

Table 3.2 Laser types and specifications

Laser type	Active material	Wavelength nm	Mode	Peak Power[a] W	Pulse length s	Coherence length cm	Efficiency %
Solid State	Al_2O_3-Cr^{+3}	694.3	Pulsed	10^8	10^{-8}	3	1
	Nd-Glass	1064	Pulsed	$5 \cdot 10^{10}$	10^{-7}	0.005	1.5–2
	Nd-YAG[b]	1064	Pulsed	10^9	10^{-9}	0.01	2-3
Color Center	LiF-F_2^+	820–1050	CW	2	–	20	40
			Pulsed	10^7	10^{-8}	–	–
Gas	He-Ne[c]	632.8	CW	0.1	–	20-30	0.01
	He-Cd	325	CW	0.08	–	3	–
		441.6	CW	0.02	–	3	–
	Ar^+ [d]	454–528.7	CW	20	–	10	0.2
			Pulsed	10^5	10^{-5}	–	–
	Excimer ArF	193	Pulsed	$8 \cdot 10^6$	10^{-8}	0.001	3
	CO_2	10600	CW	10–10^4	–	1	0.2
			Pulsed	10^5–10^6	10^{-3}–10^{-7}	–	–
Semi-conductor	GaAs	800–900	CW	0.01	–	0.01–0.1	50
			Pulsed	100	10^{-8}	–	–
Dye[e]	Rhodamine 6G in ethanol, methanol, water etc. solutions	570–620 (tunable)	CW	4	–	0.3	10-50
			Pulsed	$5 \cdot 10^7$	10^{-12}	–	–

[a] For CW lasers the mean power is given
[b] Nd-YAG lasers SS-500 delivers pulses of 1000 J; the second harmonics (530 nm) is widely used; CW operation is also possible
[c] Coherence lengths of more than 100 m have been designed; He-Ne lasers with 3 other visible lines (612, 594 and 543 nm) are available
[d] Used to pump dye lasers
[e] The shortest generated light pulses from dye lasers are 6 fs

Single mode lasers have one transverse and several spectral modes. For example, the Spectra Physics Ar^+ ion laser (Model 171) in the single mode regime has approximately 20 spectral frequencies with a coherence length of $l_{coh} \approx 20$ cm. In the single frequency regime, for $\lambda = 514.5$ nm the same laser has a coherence length of $l_{coh} > 2$ m. For applications in holography, high coherence and long time stability of gas lasers (such as He-Ne, He-Cd, Ar^+) have a strong priority.

For multimode lasers the coherence length is much smaller, for example, multimode laser diodes have coherence lengths of 100 μm. Solid state lasers and excimer lasers also have small coherence lengths and cannot be used in coherent optics. The main parameters of optical lasers are collected in Table 3.2 and Fig. 3.7).

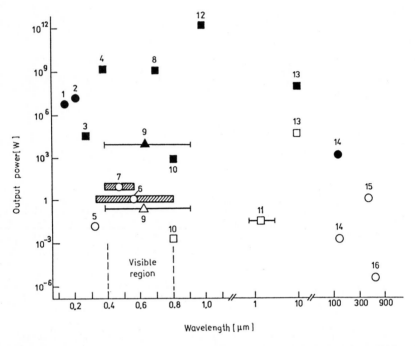

Fig. 3.7. Output power and wavelengths of common lasers: (*1*) ArF excimer (193 nm); (*2*) KrCl excimer (222 nm); (*3*) Nd-glass (4 ω_0; 266 nm); (*4*) Nd-glass ($3\omega_0$; 354.6 nm); (*5*) He-Cd (325 nm); (*6*) Kr⁺-ion (more than 17 spectral modes in the region 333.9–799.3 nm); (*7*) Ar⁺-ion (more than 10 spectral modes in the region 351.1–528.7 nm; (*8*) Al₂O₃: Cr⁺³ (694 nm); (*9*) organic dye lasers (400–900 nm); (*10*) GaAs type laser diodes (670–1550 nm); (*11*) tunable (0.8–4.0 μm) color center lasers (irradiated or additively colored LiF, NaF, NaCl etc. crystals); (*12*) Nd-YAG (1.064 μm); (*13*) CO₂ (10 μm); (*14*) water vapor (119 μm); (*15*) gaseous HCN (373 μm); (*16*) gaseous HCl (773.5 μm). Filled symbols – pulsed lasers; open symbols – continuous wave; *circles* – gas, *squares* – solid state; *triangles* – dye solution

Laser pulses (nano-, pico-, femtosecond) can be generated by different mode-locking methods with active and passive elements. The principle of mode-locking is based on the establishment of a phase relationship between spectral resonator modes which would be chaotic without mode-locking. Passive mode-locking is realized by the insertion of saturable absorbers in the resonator (e.g. dyes, or alkali halide crystals with color centers). Active mode-locking uses acousto-optic modulation of the cavity length for laser spectral mode synchronization which gives much higher output intensities. The laser pulse length depends on the switching technique [3.5,7]. However, the absolute minimum pulse length is given by the uncertainty relation [3.7]:

$$\delta v \cdot \tau \geq 1, \tag{3.16}$$

where δv is the linewidth of the emission band and τ the minimum pulse length. From (3.16) we see that the minimum pulse length is shorter for lasers with a broad emission, e.g., dye lasers (the shortest pulses generated by these lasers are 6 fs, Table 3.2).

CO_2 gas lasers can operate in the CW and pulsed regimes (in CW mode these lasers have the largest power). CO_2 lasers are used in optical communication systems, medicine (laser surgery) and machining (e.g., welding, etching). Nd-YAG lasers are also used in such situations.

Diode lasers are particularly interesting for digital optical information recording (Chap. 5). In the last 2 to 3 years, the production of diode lasers has rapidly increased and now accounts for more than 30% of the total commercial sales (98% of the total number of lasers produced) [3.8, 9]. Diode lasers are mainly used for optical memory disc systems (including CD players), which account for 75% of the total production; laser printers run second (22%) and optical communication systems third (2%).

The lasing medium in diode lasers is a semiconductor device with a p-n junction. The recombination of the injected charge carrier results in photon emission. The energy efficiency can reach more than 50%. The faces of the semiconductor (flat, parallel and polished) play the role of the end mirrors of the resonator (Fig. 3.8a). The active p-n junction, with a different refractive

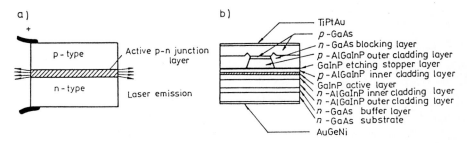

Fig. 3.8a–b. Semiconductor injection lasers – diode lasers. (**a**) The region of the p-n junction is the lasing medium, its flat, polished end surfaces are the resonator cavity mirrors. The p-n junction has a refractive index different from the p and n regions and thus acts as waveguide for the radiation. (**b**) Example of an index-guided visible range laser diode (NEC, NDL 3200; $\lambda = 670$ nm, P = 5 mW) showing the complex multilayer structure designed for maximum output power

index from the bulk material (or the multilayer semiconductor system) acts as a wave-guide for the stimulated emission. The diodes emit light from the small p-n junction region (the thickness and width of the emission region is from one to ten micrometers). The current density (up to 10^6 A/cm^2) and emitted light intensity (usually $5 \cdot 10^7$ W/cm^2) are extremely high and are limited by the thermal resistance of the semiconductor material (in the best laser diode, in single mode generation several watts from one square micrometer can be obtained) [3.10, 12]. New possibilities for optical processing and picture analysis have been introduced by quantum well (one dimensional) and quantum wire (i.e., multidimensional quantum well) lasers, discussed in the review by *Kapon* [3.11].

The parameters of commercial diode lasers are presented in Tables 3.3 and 3.4. In recent years, diodes with a high output power (or "peak power" for pulsed diode lasers) have become available, reaching 100 mW for CW operation. Single mode laser diodes with a small emission area (1-3 μm^2) have been developed for optical recording and memory systems. Diode materials, such as GaAlAs and InAlAsP, emit at shorter wavelengths which is also useful for information recording, in particular, for a higher storage density, see Sect. 2.4.4). To maximize the output power, diode lasers have a complex multilayer structure (Fig. 3.8b).

Table 3.3. Continuous wave diode laser parameters [3.12]

Manufacturer, model	Active material	Emission wavelength, linewidth (in brackets), mode structure[a] nm	Output power mW	Beam divergence degrees	Operating temperature C
NEC Corp. NDL 3200	InAlAsP	670(4); ST	4	7°30'	-10/50
NSG America LDT-780	GaAlAs	780(2); SM, MM	3	–	0/60
Spectra Diode Labs. SDL-3480-J	GaAlAs	795(5); MM	5000	10°40'	0/25
Amperex/Philips 515CQLA	GaAlAs	820(6); ML, ST	2	0.3 mrad (collimating optics)	0/60
NEC Corp. NDL 5009C	InGaAsP	1300(4); ML, ST	8	35°28'	-10/70
Mitsubishi ML-9701A	InGaAsP	1550(8);ML, ST	16	35°40'	-20/50

[a] SM – single mode; MM – multimode; ST – single transversal; SL – single longitudinal; MT – multi-transversal; ML – multi-longitudinal

Table 3.4. Pulsed diode laser parameters [3.12]

Manufacurer, model	Active material (number of elements)	Emission wavelength, linewidth (in brackets), mode structure[a] nm	Peak power W	Pulse length ns	Beam divergence[b] degrees	Operating temperature range C
Amperex/Philips 512 CQLA	GaAlAs (1)	820(4);M	250	5	30·38	-20/60
Laser Diode Inc. LD-300	GaAs (36 to 66)	904(3.5)	300 to 700	200	20	-50/70
Opto-Electronics PLS 20	InGaAsP	1550(19);S,M	0.03	0.06	25·45	15/35

[a] S - single mode; M - multimode
[b] When the beam is not radially symmetric two numbers are given

3.3 Nonlinear Optics

To discuss the nonlinear interactions of laser light in a medium, it is necessary to understand the light induced polarizability of matter. We will discuss the nonlinear optical phenomena by a simple model of an isotropic medium. The electric field of the electromagnetic light wave induces electric dipoles (in terms of electro-dynamics) or the light quanta excites the atoms (in terms of quantum mechanics) of the medium. So, elemental dipole moments parallel to a E_l are induced in the medium. The total dipole moment per unit volume, P, can be expanded as a function of E_l

$$P = X_1E_l + X_2E_l^2 + X_3E_l^3 + \dots , \tag{3.17}$$

where X_1 is the linear polarizability and X_2, X_3 are the nonlinear polarizabilities of the medium. The linear polarizability is a scalar and does not depend on the direction of propagation of the light wave in the medium nor on the direction of the electric field, E_l, of light.

Nonlinear optics is a branch of optics which describes the interaction of the light wave in a medium whose properties are changed due to different external influence (light, electric field, deformation, etc.). Rigorous analysis of nonlinear optical phenomena is possible only with tensors characterizing the properties of the medium via the direction of light propagation, polarization, light intensity etc. [3.20]. Therefore, equation (3.17) is not correct for a complete analysis of nonlinear processes in an anisotropic nonlinear material.

However, equation (3.17) allows a simple description of the main nonlinear phenomena (second harmonic generation, light modulation) used for laser spectroscopy and optical information recording. The first term in (3.17) describes the linear optical phenomena and additional terms in the expansion characterise nonlinear properties of the medium. Nonlinear polarisations are vectorial quantities and are described by tensors.[4] The polarisations of higher orders X_n decrease with increasing n (usually $X_{n+1} \ll X_n$). There are two kinds of nonlinear optical phenomena. The first leads to an optical anisotropy of the medium induced by external fields (mechanical strain, electro-magnetic). The second kind of nonlinear phenomena occurs in the medium by the interaction of light with a high intensity (and corresponding high electrical field E_l).

The first class of nonlinear optical phenomena was observed in the 19th century. In 1815 Brewster observed the photoelasticity – a birefringence in glasses induced by an external mechanical stress. In 1846 Faraday observed the rotation of the polarization plane of a light wave propagating in a transparent material under an external magnetic field. In the eighties, the electro-optic effects (i.e., electric field induced birefringence) in liquids, glasses and crystals were observed (Table 2.2). All these effects are induced by the external fields´optical anisotropy and they occur at any light intensity. The anisotropy of the medium can be used for light modulation and second and higher order harmonics generation.

The second class of nonlinear phenomena are typical for high light intensities and occur by the interaction of the light wave with the material. Such effects are nonlinear absorption, higher order harmonic interaction, stimulated light scattering etc. [3.23]. Most of these nonlinear effects were discovered after the invention of lasers, which was the birth of modern nonlinear optics.

In a medium with a definite ε and μ (dielectric constant and magnetic susceptibility), the relationship between I and E_l is :

$$E_l = \left[2I \left(\frac{\mu_0 \mu}{\varepsilon_0 \varepsilon} \right)^{1/2} \right]^{1/2} , \qquad (3.18)$$

where I is the incident light intensity in W/cm^2 and ε_0, μ_0 are the permitivity and and susceptibility of vacuum. For dielectric materials $\mu \approx 1$ and

---

[4] The correct description is given by complex functions and high order tensors [3.29]. The second order polarizability in (3.17) is then equal to $P_i^{(2)}(\omega_3 = \omega_1 + \omega_2) = 2d_{ijk}E_j(\omega_1)E_k(\omega_2)$ and is responsible for the second harmonic generation ($\omega \to 2\omega$), for sum-frequency ($\omega_3 = \omega_1 + \omega_2$) and difference-frequency ($\omega_3 = \omega_1 - \omega_2$)generation and for parametric amplification and oscillation ([3.20] p. 504).

$\sqrt{\varepsilon} = n$ giving

$$E_l \approx 27.46 \left(\frac{I}{n}\right)^{1/2}.$$ (3.19)

The intrinsic electric field in dielectrics is of the order of 10^7–10^9 V/cm, which, for a medium with a refractive index $n = 2$, means that light intensities of 10^{11} to 10^{15} W/cm^2 are required to induce nonlinear effects. Such intensities can be achieved with pulsed lasers. For typical nonlinear materials, the intrinsic atomic fields are smaller than the above mentioned values and nonlinear phenomena take place at light intensities of the order of 10^7–10^9 W/cm^2. However, for any medium, an upper limit of the critical light intensity exists at which irreversible property changes or laser damage occurs. This light intensity is called the laser damage theshold, I_d. If the light intensity in the medium exceeds the value I_d, laser damage destroy the material. Laser damage leads to macroscopic defects in the bulk or on the surface of the material. The mechanisms of laser damage are different. For instance one mechanism would be light induced heating (local melting, evaporation, thermal stimulated chemical reactions etc.). Another general mechanism in optical materials, is the electrical breakdown, i.e., an avalanche discharge in the electric field of the light wave caused by the ionization of atoms. Often both mechanisms (heating and breakdown) act together. The light induced electrical breakdown also takes place in the atmosphere and limits the light intensity in air. Atmospheric electrical breakdown occurs at field strengths of approximately 30 kV/cm, which corresponds to a light intensity of 1 MW/cm^2. This is the upper limit for light intensities of lasers in the atmosphere.

Laser damage in optical materials depends on the properties of the medium as well as from the wavelength, pulse length, laser beam diameter, temperature etc. An important role is played by inhomogeneities which increase the light scattering. Laser damage can be promoted by light-induced deformation of the material. Such a process can be induced by, for example, acoustic wave generation, photon pressure and microdefect avalanches. Usually, the surface is more sensitive for laser damage than the inside (i.e., the bulk) of the material. Therefore, all optical elements (windows, beam splitters, lenses etc.) for pulse laser technique must be covered with coatings which have a higher threshold for laser damage [3.25].

In Table 3.5, laser damage threshold intensities are given for optical materials which are important for phase matching (Sect. 3.3.3), light modulators (Sect. 3.4) and optical recording (Chap. 4). For nonlinear dielectric materials, the laser damage threshold is on the order of 10^8–10^{11} W/cm^2.

Table 3.5. Laser damage threshold intensities in dielectric and semiconductor materials [3.16, 26, 27]

Material	Threshold intensity, I_d W/cm^2	Excitation wavelength nm; pulse width ns
Silicate glass ED-2	1.10^{11}	$\lambda = 1064$ nm, $\tau = 0.125$ ns
KDP	4.10^8	$\lambda = 694.3$ nm, $\tau = 20$ ns
	2.10^{10}	$\lambda = 532$ nm, $\tau = 200$ ps
LiNbO$_3$-Fe	3.10^8	$\lambda = 1064$ nm, $\tau = 14$ ns
	3.10^9	$\lambda = 532$ nm, $\tau = 10$ ns
	9.10^{10}	$\lambda = 532$ nm, $\tau = 25$ ps
As$_{0.5}$Se$_{0.5}$ film, $d = 1$ μm	1.10^7	$\lambda = 1064$ nm, $\tau = 14$ ns
	5.10^5	$\lambda = 532$ nm, $\tau = 10$ ns
	6.10^9	$\lambda = 532$ nm, $\tau = 25$ ps
KBr	9.10^{10}	$\lambda = 532$ nm, $\tau = 25$ ps
KBr-F N$_F \leq 3 \cdot 10^{16}$ cm^{-3}	6.10^{10}	$\lambda = 532$ nm, $\tau = 25$ ps

Note, that the damage threshold increases by 1–2 orders of magnitude by a change from the nano- to picosecond pulse excitation.

For As-Se films, I_d strongly decreases upon optical excitation with a wavelength $\lambda = 532$ nm, which is near to the band gap where absorption increases. Laser damage accounts for nonlinear optical processes and optical recording, especially, by laser pulse excitation.

3.3.1 Nonlinear Absorption

One of the earliest observations of a nonlinear optical effect was the decrease in the absorption coefficient of uranyl glasses at high light intensities [3.6]. Since then, such light induced bleaching (negative absorption Sect. 3.2.1) has been observed in many materials and is even used to generate short laser pulses (passive Q-switching) [3.5,17]. The most important nonlinear absorption is the two photon process described by the third term in equation (3.17). The absorption coefficient can be written as an expansion with two terms:

$$\kappa_a = \kappa_{a1} + \kappa_{a2}, \tag{3.20}$$

where κ_{a1} is the linear term independent of the incident light intensity, whereas the two-photon absorption coefficient $k_{a2} = \beta I_0$ (β in cm/W, is a parameter of the material, Table 3.6).

From (3.4) the absorbed light intensity, I_a, is

$$I_a = I_0 - I_t = I_0 \, [1 - \exp(-\kappa_a d)], \tag{3.21}$$

where I_t is the transmitted light intensity and d the thickness of the medium. Thus, for two-photon absorption (in the case that $\kappa_a d \ll 1$) we have the following approximation:

$$I_a \approx \beta I_0^2 d, \tag{3.22}$$

that is, the absorbed intensity is proportional to the square of the incident light intensity. A dielectric requires light intensities of at least 10^7 W/cm² for nonnegligible two-photon absorption (Table 3.6).

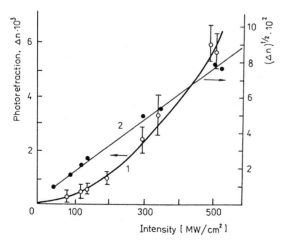

Fig. 3.9. Light induced birefringence in LiNbO₃-Cu crystals versus intensity of the second harmonic $\lambda_2 = 532$ nm of a Nd-YAG laser: curve (1) – the light induced birefringence (left y-axis); (2) $-\sqrt{\Delta n} = f(I)$ showing the two photon excitation (right y-axis) [3.15]

Table 3.6. Two-photon absorption constants [3.13, 14]

Material	Wavelength nm	$\beta \cdot 10^9$ cm/W
LiF	266	0.02
NaCl	266	3.5
KCl	266	1.7
CaF$_2$	266	0.02
GaAs	532	23
CdTe (zinc blende)	532	22
ZnS (zinc blende)	532	3.5
CdS (wurtzite)	532	5.5
As$_2$S$_3$ (crystalline)	693	E∥c 110
	693	E∥c 77
As$_2$S$_3$ (glassy)	693	142

Figure 3.9 shows an experimentally measured photorefraction in LiNbO$_3$-Cu crystals by two photon excitation. The light induced refractive Index change Δn is proportional to I_0^2 (Fig. 3.9, 2), in agreement with the two photon absorption law (3.22) [3.15]. Two-photon processes are important in a variety of optical recording methods which will be discussed in Chap. 4. Note that the small value of κ_{a2} implies that the absorption depth in two-photon processes is large, which is important for photoinduced processes in bulk materials (in the spectral region where the linear absorption is strong, see Fig. 3.1 and Table 3.1) and which means that it can be used for volume hologram recording.

3.3.2 Second Order Polarization

The second term in equation (3.17) describes various nonlinear phenomena important for laser technique and optical recording. These are the second harmonic generation (SHG), optical rectification (Sect. 3.3.3), the linear electro-optic effect (Sect. 3.4.1). All these processes are determined by the nonlinear second-order term in (3.17):

$$P^{(2)} = X_2 E_l^{(2)}, \tag{3.23}$$

and depend not only on the magnitude of E_l but also on the directionality of the second order polarizability, X_2, which is a tensor determined by the symmetry of the anisotropic medium (i.e., depends on the direction of light propagation and the direction of the electric field) [3.17–19]. For a light wave propagating in an anisotropic medium, the second order polarization of the medium is given by :

$$P_i^{(2)} = \sum_{jk} X_{ijk}^2 E_j E_k,$$ (3.24)

where E_j and E_k are the electric field components perpendicular to the direction of the light propagation and X_{ijk}^2 is the nonlinear susceptibility tensor. This can be rewritten by using the *nonlinear optical tensor d_{ijk}*

$$P_i^{(2)} = d_{ijk} E_j E_k.$$ (3.25)

The nonlinear susceptibility tensor and the nonlinear optical tensor depends on the frequency (or the wavelength) of the light wave. However, in the spectral transmission region, which is important for the second harmonic generation, the variation of the tensor elements with the wavelength is neglegible. As there is no physical difference upon an exchange of the indices in E_j and E_k in (3.25), it follows that $d_{ijk} = d_{ikj}$. We can therefore replace the subscripts kj by a single symbol according to the contraction used in crystal physics ([3.18] p. 409):

xx = 1; yy = 2; zz = 3; yz = zy = 4; xz = zx = 5; xy = yx = 6.

From this, we have six components for the electric field $E_j E_k$ (the summation over j and k for x, y, z gives 6 terms, i.e., E_x^2, E_y^2, E_z^2, $2\,E_{zy}$, $2\,E_{zx}$, $2\,E_{xy}$). We will discuss only the second order polarization $P_i^{(2)}$ and, from here on, omit the upper index. Thus, the nonlinear polarization (3.25) P_i (with three components P_x, P_y, P_z) is equal to:

$$\begin{pmatrix} P_x \\ P_y \\ P_z \end{pmatrix} = \begin{pmatrix} d_{11} & d_{12} & d_{13} & d_{14} & d_{15} & d_{16} \\ d_{21} & d_{22} & d_{23} & d_{24} & d_{25} & d_{26} \\ d_{31} & d_{32} & d_{33} & d_{34} & d_{35} & d_{36} \end{pmatrix} \begin{pmatrix} E_x^2 \\ E_y^2 \\ E_z^2 \\ 2E_z E_y \\ 2E_z E_x \\ 2E_x E_y \end{pmatrix}$$ (3.26)

The nonlinear tensor d_{kj} depends on the symmetry of the crystal. It obeys the same symmetry restrictions as the electro-optic and piezoelectric tensors. In the transparent region for nonlinear materials, d_{ij} is independent of the frequency (Table 3.7) [3.19, 20]. The forms of the tensor are given in [3.20]. GaAs crystals (cubic system, point group $\bar{4}3$ m) have a tensor d_{kj} with only three equal matrix elements:

$$\begin{pmatrix} 0\,0\,0\,d_{14}\,0\,0 \\ 0\,0\,0\,0\,d_{14}\,0 \\ 0\,0\,0\,0\,0\,d_{14} \end{pmatrix} \qquad (3.27)$$

and the components of the nonlinear polarization, according (3.26) are:

$$\begin{aligned} P_x &= 2\,d_{14}\,E_z\,E_y, \\ P_y &= 2\,d_{14}\,E_z\,E_z, \\ P_z &= 2\,d_{14}\,E_x\,E_y. \end{aligned} \qquad (3.28)$$

The directionality of the polarizability of nonlinear materials, determinated by d_{ij}, can also be used for second and higher order harmonic generation ($\omega \rightarrow m\omega$) by the interaction of the first harmonic wave with the nonlinear medium.

Table 3.7. The nonlinear optical coefficients of selected crystals [3.21]

Crystal	Point Group	λ μm	d_{ij} $1/9 \cdot 10^{-22}$ MKS
GaAs	$\bar{4}3$ m	2.12 10.6	$d_{14} = 138$ 107
LiNbO₃	3 m	1.06	$d_{31} = -4.7$ $d_{33} = -27$ $d_{22} = 3.2$
LiTaO₃	3 m	1.058	$d_{31} = -1.4$ $d_{33} = -21$ $d_{22} = 2.2$
Se	3 m	10.6	$\lvert d_{111}\rvert = 77$
SiO₂ (α-Quartz)	3 m	1.064	$d_{11} = 0.4$ $d_{14} = -0.0036$
Sr₀.₅Ba₀.₅Nb₂O₆ (SBN)	4 mm	1.064	$\lvert d_{31}\rvert = 5.3 \pm 0.5$ $\lvert d_{33}\rvert = 14.1$ $\lvert d_{15}\rvert = 7.5$
TeO₂	422	1.064	$\lvert d_{14}\rvert = 0.50$
Bi₄Ge₃O₁₂	$\bar{4}3$ m	1.064	$\lvert d_{14}\rvert = 1.6$
LiIO₃	6	1.064	$d_{31} = -5.7$ $d_{33} = -5.6$ $\lvert d_{14}\rvert = 0.25$
KH₂PO₄ (KDP)	$\bar{4}2$ m	1.318 1.06 0.6328 0.6943	$d_{36} = 0.48$ 0.50 0.57 0.56

3.3.3 Second Harmonic Generation and Phase Matching

Second (and higher) harmonic generation (SHG) is used to extend the range of laser emission for laser spectroscopy and different applications, including optical recording [3.5,11,28]. Let us return to (3.17) once more. For a sinusoidal electromagnetic field of the form $E_1 = E_0 \sin \omega t$, the light induced polarization of the medium, $P(t)$, can be written:

$$P(t) = X_1 E_0 \sin \omega t + \frac{1}{2} X_2 E_0^2 - \frac{1}{2} X_2 E_0^2 \cos 2\omega t - \frac{1}{4} X_3 E_0^3 \sin 3\omega t + ... \quad (3.29)$$

The first term in (3.29) is the first harmonic (i.e., incident light with frequency ω). The term $0.5 X_2 E_0^2$ describes the *optical rectification*, a light induced constant polarization, observed in many materials. In KDP (potassium diphosphate) crystals, for example, at $I = 10^8$ W/cm^2 the light induced permanent electrical field strength is about 5 V/cm [3.18]. This phenomenon is of theoretical interest, but, at the moment, without practical application.

The higher order terms in (3.29) describes the higher order optical harmonics, i.e., the induced light waves with frequencies 2ω, 3ω etc. At present, up to the seventh harmonic have been observed (because the polarisations are decreasing with higher orders). The second harmonic (or more generally, even number $2m\omega$ harmonics) can be induced only in anisotropic media without an inversion symmetry (not in crystals whose structure is symmetrical about a point). For high order harmonics generation, the phase matching conditions must be fulfilled. Let us analyze the optimal conditions for the second harmonic generation by a plane polarized wave propagating in a nonlinear medium in the z direction. This wave: $E_1(z,t) = E_0 \sin(k_1 z - \omega t)$ by means of the second order polarization at 2ω, i.e., $P_2(2\omega)$, induces second harmonics $E_2(z,t) = = E_0 \sin(k_2 z - 2\omega t)$, where k_1 and k_2 are the wave vector for the first and second harmonics. From (3.29) one sees that the amplitude of the second harmonic field is proportional to the square of the incident field:

$$E_0 \sin(k_2 z - 2\omega t) \sim E_0^2 \sin^2(k_1 z - \omega t). \quad (3.30)$$

Using the transformation: $\sin^2 \alpha \sim \cos 2\alpha$ for the first harmonic, we get:

$$E_0 \sin(k_2 z - 2\omega t) \sim P_2(2k_1 z - 2\omega t) \sim E_0^2 \cos(2k_1 z - 2\omega t). \quad (3.31)$$

When the phase velocities of $E_2(z, t)$ and the second order polarization $P_2(z, t)$ are equal, the maximum transfer of energy from $P_2(z, t)$ to $E_2(z, t)$ takes place and the maximum amplitude for the second harmonic is attained. This occurs when

$$k_2 - 2k_1 = 0. \qquad (3.32)$$

This is the *phase matching* requirement for second harmonic generation. In terms of the indices of refraction ($n = kc/\omega$, c is the speed of light in vacuum), equation (3.32) is related to the refractive indexes by:

$$n(2\omega) - n(\omega) = 0. \qquad (3.33)$$

In isotropic media over the normal dispersion region $n(2\omega) > n(\omega)$, and, therefore, the optimal phase matching condition (3.33) for SHG can not be achieved. In anisotropic media the magnitudes of $n(2\omega)$ and $n(\omega)$ depend on the direction of light propagation and the direction of E_l (i.e., light polarization). This can be illustrated by the index ellipsoid, or optical indicatrix shown in Fig. 3.10. The ellipsoid traces out the indices of refraction of an arbitrary anisotropic crystal in three dimensions. The three directions (x, y, z) are called the *principal dielectric axes* and n_x, n_y, n_z are the principal refractive indices of the crystal. For a given direction of light propagation, S, in the crystal, the refractive indices will be given by the major and minor axes of the ellipse, formed by intersection plane with the ellipsoid in the direction perpendicular to S. For two directions, the intersection results in a circle (in this case the two refractive indexes are equal and do not depend on the polarization direction E_l); these directions are called the *optical axes* and such substances are referred to as *biaxial crystals*. Crystals with a higher symmetry have only two principal refractive indices ($n_x = n_y$; n_z) and one optical axis (the optical indicatrix then is a rotary ellipsoid). Such crystals are called *uniaxial* or *one axis* crystals. If light propagates in the direction of the optical axis, the refractive index will not depend on the direction of E_l (light polarization), i.e., in this direction the birefringence is absent. The refractive index $n_x = n_y = n_0$ is called the ordinary and $n_z = n_e$ the extraordinary. If $n_e > n_0$, the medium is a "positive" uniaxial crystal and for $n_e < n_0$, it is "negative". Table 3.8 shows n_e and n_0 for KDP crystals as a function of wavelength.

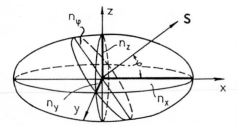

Fig. 3.10. The index ellipsoid $n(x, y, z)$ where n_x, n_y, n_z are the diagonal matrix elements of the tensor n_{ij}. For a given direction of the propagation of light S (e.g., with the angle φ to the x-axis) in an anisotropic medium, the refractive indices will be given by the major and minor axes of the ellipse formed by intersection of the ellipsoid in the direction perpendicular to S

Table 3.8. Dispersion of the refractive index in KH_2PO_4 (KDP) crystals [3.21]

Wavelength µm	Ordinary refactive index, n_o	Extraordinary refractive index, n_e
0.2000	1.622630	1.563913
0.3000	1.545570	1.498153
0.4000	1.524481	1.480244
0.5000	1.514928	1.472486
0.6000	1.509274	1.468267
0.7000	1.505235	1.465601
0.8000	1.501924	1.463708
0.9000	1.498930	1.462234
1.0000	1.496044	1.460993
1.1000	1.493147	1.459884
1.2000	1.490169	1.458845
1.3000	1.487064	1.457838
1.4000	1.483803	1.456838
1.5000	1.480363	1.455829
1.6000	1.476729	1.454797
1.7000	1.472890	1.453735
1.8000	1.468834	1.452636
1.9000	1.464555	1.451495
2.0000	1.460044	1.450308

The anisotropy of the crystal can be used to fulfill the phase matching conditions (3.33). Taking into account the dispersion, $n(\lambda)$, and the dependence of n_o and n_e on the direction of light propagation and polarization, E_l, it is possible to get $n_o(\lambda_1) = n_e(\lambda_2)$ (Fig. 3.11). The direction for phase matching θ_m is determinated by the intersection of $n_o(\lambda_1)$ and $n_e(\lambda_2)$ (Fig. 3.11). In this direction the SHG is optimal. The angle θ_m is called the phase matching angle and is equal to [3.17]

$$\sin^2 \theta_m = \frac{n_o^{-2}(\omega) - n_o^{-2}(2\omega)}{n_e^{-2}(2\omega) - n_o^{-2}(2\omega)}. \tag{3.34}$$

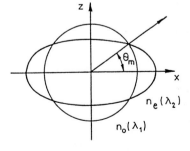

Fig. 3.11. Cross section of the index ellipsoid for the ordinary ($n_o(\lambda_1)$) and extraordinary ($n_e(\lambda_2)$) rays at different wavelengths for a uniaxial crystal. When the incident light propagates in the direction θ_m, the phase matching conditions are fulfilled ($n_o(\lambda_1) = n_e(\lambda_2)$) and, thus the second harmonic generation can occur with the maximum efficiency

In practice, condition (3.33) and (3.34) can be fulfilled only approximately, and the intensity of the second harmonic light can be estimated by [3.18]:

$$I(2\omega) \sim \frac{\omega^2}{n_m^3} \cdot d_{jk}^2 \cdot \frac{P^2(\omega)}{A} \cdot l^2 \cdot \frac{\sin^2[(k_2 - 2k_1)l/2]}{[(k_2 - 2k_1)l/2]^2}, \tag{3.35}$$

where d_{jk} is the nonlinear optical coefficient (Table 3.7); n_m is the phase matching refractive index at θ_m; l is the length of the crystal; $P(\omega)$ is the power of the initial first harmonic light and A is the area of the initial light beam. For a definite crystal, the second harmonic intensity $I(2\omega)$ is maximum when $k_2 - 2k_1 \to 0$. Note also the dependence of $I(2\omega)$ on d_{jk}. This verifies that usually SHG occurs only in anisotropic species. Typical conversion efficiencies are between $15-20\%$ at input powers of 100 MW/cm². When a nonlinear crystal is placed inside an optical resonator (a laser cavity) near 100% efficiency, SHG has been observed [3.5,18,20]. Generation of higher order harmonics $m\omega$ is difficult not only because the higher order polarizabilities X_2, X_4 are smaller, but, also, because the phase matching condition is very difficult to fulfill for $m > 3$ [3.18]. For example, it is simpler to generate the forth harmonic in two steps $(\omega \to 2\omega)$, and, then in the second step $2\omega \to 4\omega$ than directly $(\omega \to 4\omega)$. Common materials used for higher harmonic generation, among others, are KDP, ADP and LiNbO₃.

3.4 Light Modulation

Light modulation is based on nonlinear effects induced in the medium by external fields (electric, magnetic, elastic deformation). The modulation is caused by the orientation of molecules of the medium in external fields. Since the external fields can be easily varied, the optical properties can be switched with the external field frequency. This can be used to modulate the direction of propagation, polarization and intensity of light. A description of electro-optic, magneto-optic and accousto-optic modulators follows.

3.4.1 Electro-Optic Modulators

For electro-optic light modulation, electro-optic effects are used (Table 2.3). An electro-optic effect is a change in the refractive index of a medium when it is exposed to an external electrical field. This effect depends on the medium (isotropic or anisotropic), on the parameters of the external electric field, and on the direction of the polarization of light.

A general theory of electro-optic effects was developed by Pockels [3.19]. Linear (usually called the Pockels effect) and the quadratic electro-optic effects (the Kerr effect) are the most important light modulation effects.

The linear electro-optic effect can be best understood as change in the index ellipsoid of an anisotropic material in an external electric field, E. The linear electro-optic effect corresponds to second order polarization in equation (3.17). However, a correct description of the change in the refractive index of the medium in an external electrical field is only possible with the electro-optic tensor. The linear change of the refractive index, $n(x, y, z)$, in the electric field, E (x, y, z), is defined by :

$$\Delta\left(\frac{1}{n^2}\right)_i = \sum_{j=1}^{3} r_{ij} E_j \,, \tag{3.36}$$

where $(1 / n^2)_i$ has six components in the directions $x(i = 1)$, $y(i = 2)$, $z(i = 3)$, $yz = zy(i = 4)$, $xz = zx(i = 5)$, $xy = yx(i = 6)$; E_j is the external electric field with $E_1(x)$, $E_2(y)$, $E_3(z)$; and r_{ij} is the electro-optic tensor with $3 \cdot 6 = 18$ matrix elements [3.20]. Thus, (3.36) in matrix form is:

$$
\begin{pmatrix}
\Delta\left(\frac{1}{n^2}\right)_1 \\
\Delta\left(\frac{1}{n^2}\right)_2 \\
\Delta\left(\frac{1}{n^2}\right)_3 \\
\Delta\left(\frac{1}{n^2}\right)_4 \\
\Delta\left(\frac{1}{n^2}\right)_5 \\
\Delta\left(\frac{1}{n^2}\right)_6
\end{pmatrix}
=
\begin{pmatrix}
r_{11} & r_{12} & r_{13} \\
r_{21} & r_{22} & r_{23} \\
r_{31} & r_{32} & r_{33} \\
r_{41} & r_{42} & r_{43} \\
r_{51} & r_{52} & r_{53} \\
r_{61} & r_{62} & r_{63}
\end{pmatrix}
\begin{pmatrix}
E_1 \\
E_2 \\
E_3
\end{pmatrix}
\tag{3.37}
$$

The matrix elements r_{ij} are called the *electro-optic coefficients* which determine the magnitude of $\Delta(1 / n^2)$ in a definite direction of the crystal under a given field direction of E. The tensor r_{ij} has the same symmetry behavior as the nonlinear tensor, d_{ij} [3.17]. In crystals with an inversion center and in isotropic materials no linear electro-optic effect is possible. The electro-optic coefficients for some crystals used for light modulation are represented in Table 3.9.

Table 3.9. Linear electro-optic coefficients [3.21]

Substance	Symmetry	Wavelength λ μm	Electro-optic coefficients[a] r_{ik} 10^{-12} m/V		Index of refraction n_i
GaAs	$\bar{4}3$ m	0.9	$r_{41} = 1.1$		$n = 3.60$
		1.15	(T) $r_{41} = 1.43$		$n = 3.43$
		3.39	(T) $r_{41} = 1.24$		$n = 3.3$
		10.6	(T) $r_{41} = 1.51$		$n = 3.3$
LiNbO$_3$	3 m	0.633	(T) $r_{13} = 9.6$	(S) $r_{13} = 8.6$	$n_o = 2.286$
			(T) $r_{22} = 6.8$	(S) $r_{22} = 3.4$	$n_e = 2.200$
			(T) $r_{33} = 30.9$	(S) $r_{33} = 30.8$	
			(T) $r_{51} = 32.6$	(S) $r_{15} = 28$	
		1.15	(T) $r_{22} = 5.4$		$n_o = 2.229$
					$n_e = 2.150$
		3.39	(T) $r_{22} = 3.1$	(S) $r_{13} = 6.5$	$n_o = 2.136$
				(S) $r_{22} = 3.1$	$n_e = 2.073$
				(S) $r_{33} = 28$	
				(S) $r_{51} = 23$	
LiTaO$_3$	3 m	0.633	(T) $r_{13} = 8.4$	(S) $r_{13} = 7.5$	$n_o = 2.176$
			(T) $r_{22} = -0.2$	(S) $r_{22} = 1$	$n_e = 2.180$
			(T) $r_{33} = 30.5$	(S) $r_{33} = 33$	
				(S) $r_{51} = 20$	
		3.39	(S) $r_{13} = 4.5$		$n_o = 2.060$
			(S) $r_{22} = 0.3$		$n_e = 2.065$
			(S) $r_{33} = 27$		
			(S) $r_{51} = 15$		
LiIO$_3$	6	0.633	(S) $r_{13} = 4.1$	(S) $r_{33} = 6.4$	$n_o = 1.8830$
			(S) $r_{41} = 1.4$	(S) $r_{51} = 3.3$	$n_e = 1.7367$
KNbO$_3$	2 mm	0.633	(T) $r_{13} = 28$	(T) $r_{23} = 1.3$	$n_1 = 2.280$
			(T) $r_{33} = 64$	(T) $r_{51} = 105$	$n_2 = 2.329$
			(T) $r_{42} = 380$		$n_3 = 2.169$
			(S) $r_{42} = 270$		
BaTiO$_3$	4 mm	0.546	(T) $r_{51} = 1640$	(S) $r_{51} = 820$	$n_o = 2.437$
					$n_e = 2.365$
Ba$_{0.25}$Sr$_{0.75}$Nb$_2$O$_6$ (BSN)	4 mm	0.633	(T) $r_{13} = 67$	(T) $r_{51} = 42$	$n_o = 2.3117$
			(T) $r_{33} = 1340$		$n_e = 2.2987$
Bi$_{12}$GeO$_{20}$	23	0.660	(T) $r_{41} = 3.22$		$n = 2.54$
Bi$_{12}$SiO$_{20}$	23	0.633	$r_{41} = 5.0$		$n = 2.54$
KH$_2$PO$_4$ (KDP)	$\bar{4}2$ m	0.546	(T) $r_{41} = 8.77$		$n_o = 1.5115$
			(T) $r_{63} = 10.3$		$n_e = 1.4698$
		0.633	(T) $r_{41} = 8$		$n_o = 1.5074$
			(T) $r_{63} = 11$		$n_e = 1.4669$
		3.39	(T) $r_{63} = 9.7$		

[a] (T) low modulation frequency region (up to 10 kHz); (S) high modulation frequency region (~100 MHz)

The *quadratic electro-optic effect* or Kerr effect is the index ellipsoid change in an external electric field, where the induced birefringence with $\Delta n = n_\perp - n_\parallel$ is

$$\Delta n = \lambda K E^2, \tag{3.38}$$

where λ is the wavelength, K the Kerr constant, E the electric field strength and $n_\perp$, $n_\parallel$ the refractive indices perpendicular and parallel to the electric field. The Kerr effect is one of the electro-optic effects described by Pockels [3.18, 19]. It occurs in any material (solid, liquid or gas). A normally isotropic material becomes anisotropic, like a one-axis crystal with the optical axis in the direction of the external field. The Kerr effect is much larger for a material consisting of molecules with permanent dipole moments. To illustrate this, consider nonpolar benzene (C_6H_6); $K = 1.46 \cdot 10^{-10}$ cm/V^2 and, for nitrobenzene molecules ($C_6H_5NO_2$) with K = $1.4 \cdot 10^{-7}$ cm/V^2 [3.17]. The Kerr constant of glasses is usually much smaller than for liquids and varies over a wide range from 10^{-14} to 10^{-23} cm/V^2 [3.5]. The Kerr effect was the first electro-optic phenomenon used for light modulation shortly after its discovery. However, at present the linear electro-optic effect is more commonly used because the linear electro-optic coefficients are usually larger than the Kerr constant.

An *electro-optic light modulator* (an electro-optic cell) has three main elements: (1) two polarizers (Fig. 3.12); (2) an electro-optic material between them; (3) an external field. There are two different types of electro-optic cells: transverse with the direction of the electric field perpendicular to the propagation of light, and longitudinal, where the electric field is parallel to the propagation of light .

The most important application of electro-optic effects is in the modulation of the intensity of light or *amplitude modulation*. Let us analyze the electro-optic modulation by the Kerr effect. The induced birefingence by the electric field in an isotropic material according to (3.38), is equal to: $\Delta n = \lambda K E^2$. If the length of the path of light through the medium is l, then the phase difference between both waves is:

$$\Delta \varphi = \frac{2\pi}{\lambda} \cdot \lambda l K E^2 = \frac{2\pi l K U^2}{d^2}, \tag{3.39}$$

where U is the applied voltage and d is the thickness of the electro-optic material. The voltage, U_π, at which the induced phase difference is $\Delta \varphi = \pi$ is called the half-wave voltage. Such an electro-optical medium with $\Delta \varphi = \pi$ acts like a half-wave plate which changes the direction of the incident linearly polarized light [3.4]. In Fig. 3.12a, the incident light vector E_l has an angle

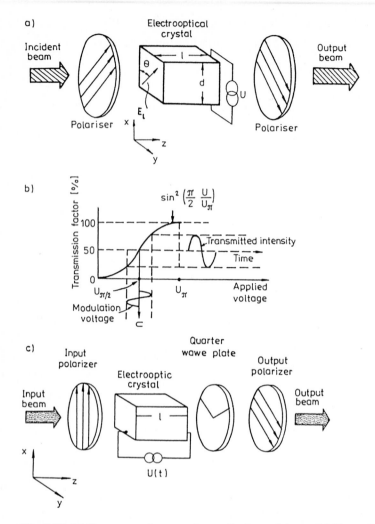

Fig. 3.12. (a) Transverse electro-optic amplitude modulator. (b) Transmission of a cross-polarized electro-optic modulator as a function of the applied voltage, $U(t)$. The modulator is biased to point $U_{\pi/2}$ which results in 50% transmission. An applied sinusoidal voltage $U(t)$ modulates the transmitted light intensity about the bias point. (c) Longitudinal electro-optic amplitude modulator, with transparent electrodes on the front and back of the electro-optic medium. $U(t)$ is biased by a fixed retardation (a quarter-wave plate) which yields 50% transmission. From [3.20]

$\theta_0 = 45°$ to the direction of the electric field and the direction E_l changes after crossing the medium by an angle of $\theta = 2\theta_0 = 90°$. By crossing input and output polarizers, the transmitted light intensity, $I(\theta)$, according Malu's law is:

$$I(\theta) = I_0 \sin^2 \theta, \tag{3.40}$$

where θ is the rotation angle of E_l caused by the electro-optic effect, I_0 is the incident light intensity and $I(\theta)$ the intensity of the output light (without reflection losses which can be eliminated using antireflection coatings).

Electro-optic amplitude modulation in the modulator is usually biased at a fixed voltage of $U_0, = 0.5U_\pi,$ which, according to (3.40), gives a transmission of 50%. A modulation voltage of $U_m(t)$, the amplitude of which is smaller than U_0, is added to modulate the output signal, $I(t)$. The small sinusoidal modulation voltage of $U_m(t) = U_0 \sin \omega_m t$ (ω_m is the modulation frequency) will change the transmitted light intensity as shown in Fig. 12b. For a given modulation voltage, U_m, the maximum of the transmitted light intensity, I, is equal to [3.17]:

$$I(t) = I_0 \sin^2\left(\frac{\pi}{4} + \frac{\pi U_m}{2U_\pi} \sin \omega_m t\right). \tag{3.41}$$

which for $U_m \ll U_\pi$ gives:

$$I(t) \approx \frac{1}{2}\left(1 + \frac{\pi U_m}{U_\pi} \sin \omega_m t\right). \tag{3.42}$$

Electro-optical modulators can also be biased by a quarter-wave plate which gives the same 50% transmission of the incident light intensity (Fig. 3.12c). In the linear electro-optic effect the birefrengence change, Δn, is determined by (3.37) and the phase difference, $\Delta \varphi$, depends on the electro-optic tensor, r_{ij} (Table 3.9). Let us examine the longitudinal electro-optical effect in KH$_2$PO$_4$(KDP) crystals with the optical axis $c \| z$ and the external field $E \| c$ (Fig. 3.12c). In the longitudinal electro-optical effect the external field strength is $E = U/l$ (l is the length of the crystal), and the induced birefringence, Δn, (and the corresponding phase difference $\Delta \varphi$) in KDP crystals with symmetry $\overline{4}2$ m, are equal to:

$$\Delta n = n_o^3 r_{63} \frac{U}{l} \ ; \ \Delta \varphi = \frac{2\pi}{\lambda} \Delta n \cdot l = \frac{2\pi n_o^3 r_{63} U}{\lambda} \ . \tag{3.43}$$

The modulated light intensity, for given U_m, can be calculated from (3.41), with the half-wave voltage determinated by:

$$U_\pi = \frac{\lambda}{2n_o^3 r_{63}}. \tag{3.44}$$

Note that the phase change (3.43) and the half-wave voltage (3.44) of the longitudinal modulator are independent of the thickness of the material. Taking r_{63} and n_o from Table 3.9, we get for $\lambda \approx 500$ nm, the magnitude of $U_\pi \approx 7.5$ kV. The half-wave voltage for some materials is shown in Table 3.10.

Table 3.10. Half-wave voltage (at $\lambda = 546$ nm) for longitudinal modulator [3.23]

Material	U_π/kV
KDP (KH$_2$PO$_4$)	7.5
KD*P (KD$_2$PO$_4$)	2.9
ADP (NH$_4$H$_2$PO$_4$)	9.2
LiNbO$_3$	4.0
LiTaO$_3$	2.7

In transversal electro-optic modulators the phase difference depends on the dimensions of the crystals (Fig. 3.12a). For KH$_2$PO$_4$ crystals with optic axis $c\|E$ and light propagating along the z direction (the light vector E_l is at $\theta = 45°$ to the x axis) the phase difference is:

$$\Delta\varphi = \frac{2\pi}{\lambda} \cdot l \left[(n_o - n_e) - \frac{n_o^3}{2} r_{63} \frac{U}{d} \right], \tag{3.45}$$

where the first term in brackets is the natural birefringence and the second is the electro-optic effect. Note, that $\Delta\varphi$ depends on both l and d. However, transverse electro-optic cells have certain advantages over longitudinal cells. Firstly, the electrodes setting up the electric field are parallel to the light beam and do not obscure it. Secondly, the field induced refractive index change, Δn, depends on the strength of the electric field in the crystal (not on the voltage between the electrodes). Low voltage Pockels cells are usually of the transverse type. High-speed Pockels cells, which require low capacitance and, therefore, small electrodes, are often of the longitudinal type.

For short time switching, electric fields with very high frequencies are necessary. A limitation on the modulation periods, τ_m, is the transit time of the light wave through the crystal. For KDP crystals ($n \approx 1.5$) with length $l = 1$ cm, the transit time is $\tau_t \approx 50$ ps. For an effective modulation, τ_m must be much longer than the transit time. Therefore, the modulation period for KDP crystals can not be shorter than 0.1 ns [3.20]. In liquids, the Kerr effect allows a modulation time of the order of 1 ps (in CS$_2$ $\tau = 2$ ps and in C$_6$H$_5$NO$_2$ $\tau = 50$ ps). Shorter times (of the order 10^{-14} s) are observed in crystals with electrically induced polarization (this is possible only in complicate devices). A more detailed analysis of electro-optic materials is given in [3.17,18,20].

3.4.2 Magneto-Optic Modulators

Magneto-optic modulation is based on the Faraday and magnetic Kerr effects, i.e., optical activity of medium induced by an external magnetic field. The Faraday effect entails the rotation of the plane of polarised ight in a transparent active medium in a magnetic field. This effect was observed by Faraday in liquids and glasses in 1846 and was the first phenomenon which demonstrated the interaction of an electromagnetic field with materials. Another magneto-optic effect which is analogous to the Kerr electro-optic birefringence was observed by Cotton and Mouton in 1905[5] [3.4]. However, the Cotton-Mouton constants for various materials are small. Therefore, this effect has no practical applications.

The Faraday rotation is used to make light modulators and magneto-optic recordings (Sect. 5.2.4). The angle by which the direction of polarization is changed, is given by:

$$\theta = \rho_v \cdot l \cdot H, \tag{3.46}$$

where ρ_v is the Verdet constant, l is the length of the sample and H is the magnetic field strength parallel to the direction of light propagation. The Verdet constant for liquids and glasses is small, but much larger in ferromagnetics and antiferromagnetics (Table 3.11).

The *magneto-optic Kerr effect* is the conversion of linearly polarized light into elliptically polarized light upon reflection from ferromagnetic or antiferro-magnetic surfaces in an external magnetic field.

Magneto-optic material is placed between two crossed polarizers to create a magneto-optic light modulator. However, for light modulation the electro-optic effect and the acousto-optic effect are more common. The most important application of the magneto-optic effect is recording in ferrimagnetics (Sect. 5.2.4).

---

[5] There was an early observation of this effect by Kerr in 1901 but the systematic investigations were done by *Cotton* and *Mouton*.

Table 3.11. Verdet constant for selected materials [3.23]

Material	Wavelength μm	Verdet constant* min/(Oe·cm)
CS_2	0.5893	0.02
H_2O	0.5893	0.013
α-SiO_2 (quartz)	0.5893	0.01664
C_6H_6	0.5893	0.03
a-SiO_2 (fused quartz)	0.5893	0.004
ZnS	0.666	0.234
GaAs	1.06	0.3
Si	15	0.1
$Y_3Fe_5O_{12}$	1–2	300
$CrBr_3$	0.5	1600

*To convert Oe cm to MKS units [min · m^{-1} · (A · m^{-1})$^{-1}$] multiply the above numbers by 1.257

3.4.3 Acousto-Optic Modulators

Acousto-optic modulators use a pressure deformation of a medium to alter its optical indicatrix. The history of this phenomenon is long. *Brewster* discovered in 1815 that an applied mechanical stress could modify the index ellipsoid and this behaviour is known as *mechanical birefringence* or *photoelasticity*. Acousto-optic modulation was suggested in 1921 by *Brillouin* who developed a theory of photoelasticity in terms of the light diffraction from transient gratings created by acoustic waves in the medium. *Debye* and *Sears*, and independently, *Biquard*, observed the diffraction of light by sound waves in 1932 [3.23].

The acousto-optic modulation leads to a change in the refractive index of a material induced by a sound wave. If an acoustic wave with the frequency f and velocity v_a propagates through an optically transparent medium in the direction z, the change in the refractive index $\Delta n(z, t)$ is equal to:

$$\Delta n(z, t) = \Delta n \sin(k_a z - 2\pi f \cdot t), \tag{3.47}$$

where Δn is the magnitude of the acoustic wave induced refractive index, $k_a = 2\pi/\lambda_a = 2\pi f/v_a$ is the magnitude of the acoustic wave vector, and f is the acoustic wave frequency. The magnitude of Δn depends on the acoustic strain ε_a (i.e., the acoustic wave induced local deformation; ε_a is a dimensionless parameter). The acousto-optic change in the refractive index is equal to:

$$\Delta n = -\frac{n^3 p}{2} \varepsilon_a \, , \qquad\qquad (3.48)$$

where n is the refractive index of the material and p is the photoelastic constant (the parameter p is also dimensionless).

The acousto-optic effect is a nonlinear optical phenomenon which in an isotropic optical medium induces optical anisotropy, i.e., birefringence. Thus, an isotropic material is transformed into an optical anisotropic medium (like an uniaxial crystal) with two different refractive indices. Therefore, for any isotropic material, two photoelastic constants (strain-optic-coefficients [3.20]) determinate the acousto-optical properties. However, in some isotropic materials the two acousto-optical constants are equal and the change in the refractive index, is determinated by (3.48) independent of the direction in the material. The parameters of such simple isotropic materials are tabulated in Table 3.12.

In anisotropic materials the acousto-optic effect leads to a change in the crystal symmetry with a corresponding increase in the anisotropy of the material. Thus, an uniaxial crystal in an acoustic strain field becomes a biaxial and a biaxial crystal becomes more anisotropic (i.e., is transformed into a material with lower symmetry). In crystalline materials the acousto-optic effect is determined by a strain-optic-tensor of the fourth rank ([3.20] p. 318–329).

Table 3.12. Charateristics of selected acousto-optic materials at $\lambda = 633$ nm [3.19, 21]

Materials	Spectral transmission µm	Density g/cm^3	Refractive index, n	Acoustic wave velocity 10^5 cm/s	Photoelastic constant, p	Relative diffraction constant, M_r
Water, H_2O	0.25–1.0	1.00	1.33	1.50	0.31	1.00[b]
Quartz glass	0.2–5	2.20	1.457	5.96	0.20	0.006
Flintglass, (SF-4)	0.4–1.8	3.59	1.616	3.63	0.25	0.12
Polystyrene	0.4–3.0	1.06	1.59	2.35	0.31	0.8
a-As_2S_3	0.6–11.0	3.20	2.63	2.60	–	–
a-Se	1.0–20.0	4.27	2.50[a]	1.83	–	–
$PbMoO_4$	0.4–5.0	6.95	2.30	3.75	0.28	0.22
$LiNbO_3$	0.4–4.5	4.7	2.20	7.40	0.15	0.012
Al_2O_3	0.13–5.0	4.0	1.76	11.00	0.17	0.001
LiF	0.1–8.0	2.6	1.39	6.00	0.13	0.001

[a] At $\lambda = 1153$ nm

[b] The magnitude of M for water is equal to $M_o = (n^6 p^2)/(\rho v_a^3) = 1.576 \cdot 10^{-11}$ m$^2 \cdot$W^{-1}
$M_r = M_i/M_o$, where M_i is the magnitude for a given material

Crystals exemplify an acousto-optical effect that is dependent on the direction of the wave of light, the field of the acoustic strain and the plane of polarisation of the light. However, in such a situation the problem can be simplified in such a way that only one tensor component is important for a definite direction in the anisotropic material. Thus, the simple approximation (3.48) can also be used for a crystal. This can be implemented, e.g., for $PbMoO_4$ crystals where along the **c** axis the acousto-optic properties are determined by one strain-optic-coefficient: $p_{33} = 0.30$.

Acoustic waves, according to (3.47), induce a moving phase grating in the medium with a period equal to the acoustic wavelength $\Lambda = \lambda_a$ (Fig. 3.13). For light modulation or light beam deflection, the Bragg diffraction from the acoustic phase grating is used (Fig. 3.13). Such acoustic phase grating has the same properties as a light induced holographic phase grating (Chap. 2, formula (2.7)). For acousto-optic light modulation, the diffraction of the light beam is used. The diffraction takes place under the Bragg angle, according to (2.7) on the grating with period $\Lambda = \lambda_a$

$$\sin \theta = \frac{\lambda}{2n\lambda_a} = \frac{\lambda \cdot f}{2n v_a} \, , \qquad\qquad (3.49)$$

where λ is the wavelength of light in a vacuum, n is the refractive index of the medium, v_a is the velocity of sound, f is the acoustic frequency, and $\lambda_a = v_a/f$ is the acoustic wavelength. The angle of deflection (i.e., the difference between the direction of the initial and the wave diffracted), $\Delta\theta$, is equal to $\Delta\theta = 2\theta$ (Fig. 3.13). The Bragg angle (3.49) depends on the wavelength of light and the grating period $\Lambda = \lambda_a$, which is reciprocal to the acoustic frequency. The Bragg angle is changed, according to (3.49), upon a change in the frequency. Such simple considerations are the basic principles of acousto-optic light modulation.

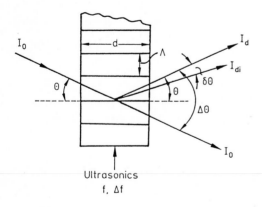

Fig. 3.13. A typical acousto-optic deflector for light modulation: d – thickness of the acousto-optic material; θ – the Bragg angle; $\Delta\theta = 2\theta$ – the angle of deflection ; $\delta\theta$ – the smallest resolvable angle of deflection; I_0 – the incident light beam; I'_0 – the nondiffracted transmitted light beam; I_d, I_{di} – diffracted light beams; Λ – the period of the phase grating; λ_a – the acoustic wavelength ($\Lambda = \lambda_a$)

High frequencies from 100 MHz to 1 GHz are used in acousto-optic modulators. However, the acoustic wavelength, λ_a, at these frequencies is much larger than the wavelength of light (in PbMoO$_4$ at $f = 500$ MHz the wavelength of sound is $\lambda_a = 7.5$ μm, Table 3.12). Therefore, the angles of diffraction according (3.49), are only several degrees (e.g., for PbMoO$_4$ at $f = 5.10^8$ Hz; $\theta \approx 1°$ and for water at the same frequency $\theta \approx 4.5°$). The width of the light beam must be also accounted in deflection experiments. The distribution of light in lasers is of a Gaussian nature and the angle of divergence, θ_{div}, is determined by equation (3.14) (Sect. 3.2). For a He-Ne laser with the Gaussian beam radius of $w_0 = 1$ mm the divergence angle $\theta_{div} \approx 0.01°$. Such light beam divergence ($\pm \theta_{div}$) determines the angular resolution of the acousto-optic modulator. The acousto-optic deflection must be greater than the minimal spatial resolution angle $\delta\theta \approx 2\theta_{div}$. Thus the maximal deflection angle is $\Delta\theta = 2\theta$, the maximal number of deflected light beams N is equal to $N \approx \Delta\theta/\delta\theta$. Thus gives for the PbMoO$_4$ acousto-optic medium (at $f = 5.10^8$ Hz; $\theta \approx 1°$) the value $N = \Delta\theta/\delta\theta \approx 40$.

The diffraction efficiency of an acousto-optic modulator can be estimated with the equation for a 3-D phase hologram. The diffraction efficiency is a sinus-function of the magnitude of the light induced refractive index (Chap. 2, Table 2.5). According (3.48) Δn is proportional to the acoustic strain ε_a. It can be shown that the magnitude of ε_a is equal to

$$\varepsilon_a = \left(\frac{2I_a}{\rho v_a^3} \right)^{1/2}, \tag{3.50}$$

where I_a is the acoustic intensity [W·m^{-2}], ρ is the density [kg·m^{-3}], and v_a is the acoustic wave velocity [m·s^{-1}] ([3.17] p. 346). According (3.48) and (3.50) the diffraction efficiency (Chap. 2) is equal to

$$\eta = \sin^2\left(\frac{\pi d}{\lambda} |\Delta n| \right) = \sin^2\left[\frac{\pi d}{\lambda} \cdot \left(\frac{n^6 p^2 I_a}{2\rho v_a^3} \right)^{1/2} \right], \tag{3.51}$$

where d is the thickness of the acousto-optic material and I_a is the acoustic intensity (the other parameters were explained above). The diffraction efficiency of an acousto-optic modulator according (3.51) is determinated by the term under the square root. Therefore, the diffraction efficiency of various materials can be described by the diffraction constant M equal to

$$M = \frac{n^6 p^2}{\rho v_a^3}. \tag{3.52}$$

The dimension of M is [m^2·W^{-1}]. For acousto-optic material characteristic a relative constant $M_r = M_i/M_o$ is used, where M_i is the constant for a definite material and M_o is the diffraction constant for water choosed as an etalon material (Table 3.12). Thus the equation (3.51) can be simplified for any acousto-optic material

$$\eta = \sin^2\left[\frac{\pi d}{\lambda}(0.5\ MI_a)^{1/2}\right]. \tag{3.53}$$

For water (Table 3.12 and 3.52) and the light wavelength $\lambda = 633$ nm, equation (3.53) can be simplified to:

$$\eta = \sin^2\left(1.4\ d\sqrt{I_a}\right), \tag{3.54}$$

where d is measured in m and I_a in W·m^{-2}. For different materials and wavelengths we get:

$$\eta = \sin^2\left(\frac{886}{\lambda} \cdot d \cdot \sqrt{M_r I_a}\right), \tag{3.55}$$

where λ is the wavelength of the modulated light (in nm) and M_r is the relative diffraction constant at $\lambda = 633$ nm (Table 3.12).

Equations (3.54) and (3.55) are useful for diffraction efficiency estimations in acousto-optic devices: we get for water at $\lambda = 633$ nm, $I_a = 10^5$ W·m^{-2} and $d = 1$ mm, the diffraction efficiency: $\eta \approx 18\%$; for PbMoO$_3$ at $\lambda = 514$ nm, $d = 1$ mm (10^{-3} m) and $I_a = 10^6$ W·m^{-2} the diffraction efficiency is: $\eta \approx 52\%$.

The acousto-optic modulator as well as the electro-optic modulator are nonlinear optical devices, and the linearity is established in the neighborhood of a transmittance of 50%.

Acousto-optic devices are used for beam deflections as well as for light modulation. An acousto-optic light modulator is an acousto-optic beam deflector followed by a lens and a pinhole (such spatial filtering system rejects the undeflected beam and passes the deflected, Fig. 3.13). The modulation takes place by moving acoustic waves through the acousto-optic material. Thus, the light beam is scattered by the vibrating acoustic wave grating, a Doppler shift in the frequency of light v takes place, i.e., $v \pm f$, where f is the acoustic wave frequency. However, the Doppler shift is small because $f \ll v$.

Parameters of Acousto-Optic Modulators. As mentioned above, the main parameter of the acousto-optic material is the photoelastic constant, p (or the strain-optic-coefficients of the tensor for crystals). The diffraction efficiency is determinated by (3.53–55), which, for a PbMoO$_4$ crystalline modulator,

($\lambda = 0.633$ μm; $d = 1$ mm; $I_a = 100$ W/cm^2; acoustic beam cross section is 1 mm^2) gives $\eta = 40\%$.

The bandwidth of the acousto-optic modulator, Δf, is determinated by the angular spread of the light beam (i.e., $\delta\theta = 2\theta_{div}$). Usually, the maximum bandwidth is approximately one-half the acoustic frequency, i.e., $\Delta f = 0.5f$. This means that large modulation is possible only for high-frequency induced acoustic phase gratings (i.e., high frequency Bragg diffraction). On the otherhand, in the frequency range $f \pm \Delta f$, the deflected beam (on the Bragg angle θ) does not interfere with the undiffracted, i.e., the angular spread of the light beam $\delta\theta = 2\theta_{div}$ must be smaller than the Bragg angle ($\delta\theta < \theta$). Thus, θ_{div} depends on the light beam width (equation (3.14)) and the diffraction angle, θ, depends on the acoustic frequency, the relation: $\delta\theta \leq \theta$ also determines the magnitude of the acoustic bandwidth Δf.

3.5 Photoelectric Detectors

Detectors of light, including ultraviolet (UV) and infrared (IR) radiation, fall into two classes: quantum detectors and thermal detectors. *Quantum detectors* are fast response devices which count the number of quanta per unit time that interact with the detector. *Thermal detectors* are slower. The absorbed lightquanta heat the detector element and the ensuing temperature rise is measured via radiation thermocouples, bolometers, or pyroelectric detectors. Quantum detectors are practically always used in optoelectronics and optical recording. These detectors can be further classified, i.e. photoconducting, or photovoltaic, and photoemissive devices.

Photoemission devices are based on light induced emission of electrons. The simplest phototube is a vacuum diode consisting of a photocathode and an anode in an evacuated glass envelope. The photocathode is a material with high quantum efficiency over a given spectral region. Usually, the phototube is combined with a secondary electron emitting chain of dynodes which produces (in an electric field from a voltage of about 100 V between dynodes) an amplification of the initial photoelectrons by a factor of 5 to 10 for each dynode. Such a system is called a photomultiplier, and can have as many as 12 or more secondary dynodes. The overall gain in the photomultiplier may be up to 10^{10}. The speed of response of the photomultiplier is limited by the transit time of electrons through the tube, which is greater than that of the vacuum photodiode and not shorter than 10 ns.

Silicon *photovoltaic detectors* (solar cells) are widely used in solar energy conversion. Such photovoltaic detectors are semiconductor diodes with p-n junctions. Under illumination, electrons and holes at the p-n junction are separated and generate a voltage. Higher amplification of the primary signal can be obtained with an avalanche photodiode, which can be quite fast and sensitive. Such photodetectors have a high quantum efficiency (Table. 3.13).

Table 3.13. Characteristics for photocathodes (for photomultipliers) and vacuum photo-diodes from RCA [3.12]

Designa-tion	Photo-sensitive material	Type of sensor	Wavelength of maximum re-sponse, λ_{max} nm	Radiant response at λ_{max} mA/W	Quantum efficiency at λ_{max} %	Photocathode dark emission at 25 °C fA/cm^2
S-4	Cs-Sb	PE[a]	400	40	12.4	0.2
S-5	Cs-Sb	PE	340	50	18.2	0.3
S-10	Ag-Bi-O-Cs	PE	450	20	5.5	70
S-14	Ge	p-n[b]	1500	520[e]	43[e]	–
S-23	Rb-Te	PE	240	4	2	0.001
	GaAs	PE	830	68	10	0.1
	Ga-As-P	PE	400	45	14	0.01
	Si	PV[c]	860	580[f]	83.5[f]	–
	Si	PC[d]	900	620[g]	85[g]	–

[a] PE-photoemitter [b] p-n junction [c] PV-photovoltaic [d] PC-photoconductor
[e] With 45 V polarizing voltage [f] Photovoltaic short-circuit response
[g] For a wafer thickness of approximately 150 μm

Most detectors in use today, particularly in the infrared and visible region, are based on *photoconductivity*, that is, changes in the resistance upon irradiation. Intrinsic photoconductivity requires the excitation of a free electron-hole pair by an incident photon whose energy is at least as great as the energy gap of the photoconductor $h\nu \geq E_g$. The long wavelength limit of an intrinsic photoconductor, therefore, is equal to:

$$\lambda_g = \frac{hc}{E_g} \approx \frac{1240}{E_g} \quad , \tag{3.56}$$

with λ_g in nm and E_g in eV, No intrinsic photoconductivity occurs at wavelengths greater than λ_g; with values listed in Table 3.14 for several materials. Extrinsic photoconductivity, however, can occur when an incident photon with $h\nu \leq E_g$ produces an electron-hole pair at an impurity center.

For the infrared region $\lambda \geq 3$ μm, the photoconductor must be cooled to suppress the spontaneous dark electron-hole pair formation. The steady state intrinsic photocurrent is:

Table 3.14. Energy band gaps E_g and λ_g for intrinsic photoconductivity

Material	Temperature K	E_g eV	λ_g µm
CdS	295	2.4	0.52
CdSe	295	1.8	0.69
CdTe	295	1.50	0.83
GaP	295	2.24	0.55
GaAs	295	1.35	0.92
Si	295	1.12	1.1
Ge	295	0.67	1.8
PbS	295	0.42	2.9
PbSe	195	0.23	5.4
InAs	195	0.39	3.2
InSb	77	0.23	5.4
$Pb_{0.2}Sn_{0.8}Te$	77	0.10	12
$Hg_{0.8}Cd_{0.2}Te$	77	0.10	12

$$i_S = q\eta N_\lambda G, \tag{3.57}$$

where η is the quantum efficiency (i.e., the number of incident photons required to induce a single electron-hole pair formation); q is the electronic charge; N_λ is the number of photons with wavelength λ absorbed in the semiconductor; and G is the gain – the total number of electrons which are produced at the output of the detector through all intermediate components for each absorbed photon. The photoconductivity gain can also be expressed as the ratio of the free carrier lifetime, τ, (i.e., the lifetime of an electron-hole pair before it recombines) to the transit time, τ_0, of the electrons between the electrodes separated by a distance l, so that:

$$G = \frac{\tau}{\tau_0}, \tag{3.58}$$

$$\tau_0 = \frac{l^2}{\mu U_A}, \tag{3.59}$$

where μ is the charge carrier mobility and U_A is the bias voltage. Taking into account (3.58, 59), we get from (3.57):

$$i_S = \frac{q\eta N_\lambda \mu \tau U_A}{l^2}. \tag{3.60}$$

The number of photons absorbed is related to the absorbed monochromatic power (P_λ), $N_\lambda = P_\lambda \cdot \lambda / hc$, thus

$$i_S = \frac{q\eta P_\lambda \lambda \mu \tau U_A}{hcl^2}.$$
(3.61)

From (3.61) the photocurrent is proportional to the absorbed power of light and depends linearly on the wavelength (for $\lambda \leq \lambda_g$). The speed of response of a detector is determined by the free carrier lifetime, τ. It sets the limit on the maximum short circuit current and output voltage that the detector can measure accurately with a given modulated optical input signal. At high input signal frequencies, if the detector is too slow, the device might not detect all the incoming photons and the photocurrent kinetics $i(t)$ does not describe the light signal $I(t)$. The shortest relaxation time for photoconducting detectors is of the order of $\tau \approx 50$ ps, corresponding to a frequency $f \approx 20$ GHz.

Detector Performance Parameters. The performance of a detector is evaluated primarily on the basis of three main parameters: the response of the detector (i.e., the output for a given irradiance which is related to the minimum detectable input power); the spectral sensitivity; and the modulation frequency response.

The response of a detector R is the ratio of the detector output to the input power. This is a function of the gain of the device, the minimum detectable signal, and the quantum efficiency. The precise definition depends upon the application. In photodetectors the response is usually given in volts or amperes per watt (or in more practical units, microvolts or microamperes per microwatt, Fig. 3.14).

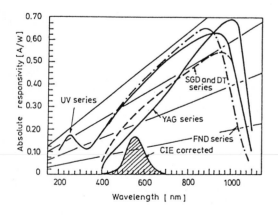

Fig. 3.14. Absolute spectral response of various silicon photo-diodes (the straight lines correspond to constant quantum efficiency). From [3.12]

The response at a given wavelength is called the *spectral response*. The spectral response also depends on the external components of the device as well as on the photosensitive medium. For example, Si photodiodes can be adapted to UV detection by the addition of transparent CaF_2 windows. The quantum efficiency, for most photodetectors is of the order of 10%, although some silicon detectors have quantum efficiencies close to 100% (Table 3.13).

Another important parameter already discussed above, is the output of the detector as a function of the input signal frequency. At high input signal modulation frequencies, if the detector is too slow, the output signal is distorted. The speed of response may be limited by various phenomena. Take for example, vacuum tube photoemission devices where the limiting factor is the transit time of the electrons from one electrode to the next. The transit time of the charge carriers in photoconductors is determined by the distance between the electrodes and the charge carrier mobility in the semiconductor (retrapping processes).

All electrical systems carry noise, due to electrical or thermal fluctuations. The easiest way to reduce the noise is simply to cool the photodetector (to liquid nitrogen or liquid helium temperatures). This surpresses the spurious emission of electrons from the dynodes (dark current). However, this is usually important only for photodetectors in the infrared spectral region.

4. Photoinduced Processes in Optical Recording Materials

In most materials the recording process is a charge transfer reaction with a change in the microstructure or a local phase transition, leading to a stable photochromic (light induced change in the coefficient of absorbtion) or photorefractive effect (light induced change in the refractive index). Both of these effects are usually diffraction limited, i.e., the minimum size of the recording area can not be smaller than the wavelength ($d_{min} \geq \lambda$). These limitations are overcome in optical recording materials with zero-phonon lines (spectral hole burning) [4.1–3].

In this chapter, we will discuss the photoinduced reactions of real time optical recording in different materials. As mentioned in Chap. 2, optical recordings can be digital or analog (e.g.,holographic). At present, only a few real time recording materials are commercially available (photothermoplasts, optical memory discs (Chap. 5)). For many applications new materials are necessary which can be easily technologically produced. Therefore, limitations of the physico-chemical and recording properties of media, as well as the principles of manufacturing, are examined here.

4.1 Inorganic Photochromic Crystals

The most widely investigated photochromic materials, those in which upon illumination $\Delta k \gg \Delta n$, are alkali halide crystals. These materials are transparent in the ultraviolet and visible spectral region (Fig. 4.1). Therefore, optical recording with visible or ultraviolet light in these crystals is only possible on defects, having intrinsic or impurity absorbtion bands in this spectral region.

Alkali halide crystals are ionically bound and the crystal lattice is an alternating network of cations and anions. The crystal defects which are important for optical recording are illustrated in Fig. 4.1. These are single

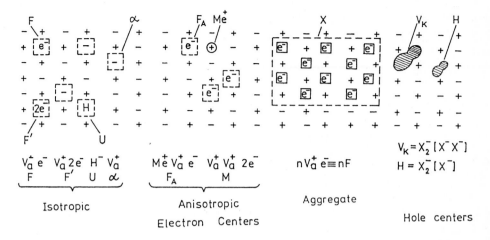

Fig. 4.1. Types of defects in alkali halide crystals: V_a, V_c - anion and cation vacancy. An anion vacancy filled with an electron is an F-center, an F'-center is where two electrons are captured by an anion vacancy. Me^+ is an impurity atom substituting a cation of the lattice; an F_A center is an F-center with a neighboring impurity Me^+; an M-center is formed by two neighboring F-centers; an X-center is a cluster of F-centers with a quasi-metallic bond; U denotes a hydrogen atom which substitutes a halogen atom; H is an interstitial X_2^- anion substituting the regular anion (the H-center is neutral relative to the lattice); V_k is an X_2^- molecule substituting two neighboring anion cites in the lattice (the V_k-center is positively charged relative to the lattice)

lattice defects (color centers or impurity atoms), which occupy one lattice position, and multiple electron color centers (M-, X-centers). Electron color centers are more useful for optical information storage than localized holes (H-, V_k-centers) because these centers are stable at room temperature and their absorption spectra are in the visible region, whereas, most hole centers absorb in the ultraviolet and are thermally annealed at room temperature. Impurity defects are also used in optical recording (e.g., rare earth element impurity defects in CaF_2 crystals).

For photochromic information recording in crystals, only certain defect states can be used. They must be stable at room temperature before and after recording and, ideally, have a strong absorption in the visible spectral region (this is necessary for high efficiency in recording). Furthermore, the light sensitivity (i.e., the quantum efficiency, Chap. 2) must be high, as also, the signal-to-noise ratio (SNR). Therefore, large absorption coefficient changes are necessary, without subsequent thermal annealing. For reversible optical recording, the possibility of optical or thermal erasing is an additional condition. These limitations restrict the many known photoinduced processes in ionic crystals to a select group. There is also a size limitation on crystalline recording materials: it is difficult to make large (with a size more than 10 cm) homogeneous samples. Therefore, glassy materials, especially thin films, are preferred over crystalline recording materials (these materials can be easily produced and have a more homogeneous macroscopic structure for optical recording, Sect. 4.3 and Chap. 5) [4.4].

Defects in crystalline materials can be produced in different ways. Impurity defects are usually created during the crystal growth process or by thermal diffusion from the vapor or solid phase [4.5, 6]. Color centers (intrinsic defects) in ionic crystals can be generated by irradiation (with X-, γ-rays or fast particles) [4.7, 8, 9] or by "additive coloration" [4.10]. Irradiance with γ- or X-rays creates equal concentrations of various electron and hole centers which, unfortunately, mostly can not be used for optical recording. This is because firstly, the absorption bands of different color centers overlap, preventing efficient recording and reading at a fixed wavelength, Secondly, the thermal instability of various color centers leads to their decomposition (annealing, aggregation, chemical reactions). Therefore, irradiated crystals are used for optical information storage only in special applications. On the other hand, additive coloration is commonly used to produce stable color centers in alkali halide and other ionic crystals [4.10, 11]. This is a thermal treatment of the crystal, in alkali metal vapor at temperatures ($T \approx 0.7\ T_m$) close to the melting point. Atom and electron diffusion creates nonstoichiometric electronic defects, i.e., only electron color centers and their aggregates, without hole centers.

The main photoinduced reaction used for optical recording in alkali halide and other ionic crystals is a charge transfer between point defects (for example, F $\Leftrightarrow$ F'), aggregation of single color centers or point defects (e.g., F-center aggregation to X-centers) and the orientation of anisotropic color centers (M-center). Besides these electron transport processes, light induced atom diffusion and aggregation are also used. However, as mentioned above, many light induced reactions in alkali halide crystals cannot be used for real time recording at room temperature. For example, the light induced charge transfer between an F- and F'-center is possible only at low temperature. The same limitations also exist for light induced orientation of anisotropic F_A-centers or F $\Leftrightarrow$ U reactions (Fig. 4.1). Another hindering factor is the low light sensitivity of many real time photochromic reactions. Therefore, here, we will discuss only the most important photochromic reactions in simple ionic crystals: charge transfer between impurity centers; the anisotropic photochromic effect and the aggregation of F-centers.

4.1.1 Light Induced Charge Transfer Reactions

Photoinduced charge transfer between impurity defects in CaF_2-Eu, Sm crystals have been tested for use as a means of information storage [4.12]. The impurity centers: Eu^{2+} and Sm^{3+}, acting as electron traps, have different absorption bands with maxima at 255 and 310 nm. The recording is written by photoexcitation of the Eu^{2+}-center at λ_1, with subsequent charge transfer to the Sm^{3+}-center:

$$Eu^{2+} + Sm^{3+} \xrightarrow{\lambda_1} Eu^{3+} + Sm^{2+}. \qquad (4.1)$$

The information can be read from both, the Eu^{3+} or Sm^{2+} centers , i.e., two channel readings are possible. Optical erasure – by irradiation at λ_2 in the Sm^{2+} absorption peak, and thermal erasure, which promotes the electrons out of the Eu^{3+} and Sm^{2+}, traps and leads back to the initial state Eu^{2+}, Sm^{3+}, are possible. Similar reactions take place in other impurity doped crystals. However, the low light sensitivity ($S^{-1} \approx 1$ J/cm^2) limits the application of such systems.

4.1.2 Anisotropic Photochromic Effect

The anisotropic photochromic effect for making real time optical recordings is the reorientation of an M-center by light, i.e., a double F-center, in alkali halides. The mechanism of these reactions in KCl and NaF crystals was elucidated by *Schneider* [4.8, 9]. This is a complex process occurring at room temperature (Fig. 4.2). In the first step, an absorbed photon from a linearly polarized laser induces ionization of an M-center, $M + h\nu \rightarrow M^+ + e^-$. Since the light is polarized, only M-centers which are oriented parallel to the direction of polarization, E_l, will be ionized. The free electrons are captured by F-centers, F $+ e^- \rightarrow F'$. Because the F'-centers are not stable at room temperature, they relax via $F' \rightarrow F + e^-$. The thermally released electron recombines with a M^+-center, $M^+ + e \rightarrow M$, and the dipole of the M-center is oriented in the direction of E_l. This light induced two step ionization-recombination reaction occurs within nanoseconds. Such a recording can be read using two light beams $I_1(\lambda_M)$ and $I_2(\lambda_{F'})$, where λ_M and $\lambda_{F'}$ are the absorption wavelengths of the M- and F'-centers. The first light beam, I_1, again induces photoionization of an M-center, but capture of the emitted electron by an F-center to make an F' is prevented by the second beam, I_2. Hence, recombination of the ionized M-center occurs immediately and the medium remains intact. Information can be optically erased by irradiating the crystal with unpolarized light in the M-center absorption band, i.e., only $I_1(\lambda_M)$.

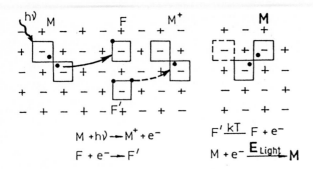

$$M + h\nu \longrightarrow M^+ + e^-$$

$$F + e^- \longrightarrow F'$$

$$F' \xrightarrow{kT} F + e^-$$

$$M + e^- \xrightarrow{E\ Light} M$$

Fig. 4.2. Light induced reorientation of M-centers in alkali halides. *Plusses* and *minuses* denote cation and anion lattice sites, the *dots* represent electrons. The recording takes place in four steps: (*1*) polarized light ionizes the M-center: $M + h\nu \rightarrow M^+ + e^-$; (*2*) the electron is captured by an F-center which leads to a thermaly unstable F'-center; (*3*) the F'-center is thermaly ionized; (*4*) the recombination of the electron with the M^+-center in the electric field of the polarized light reorientates the M-center

Table 4.1. Light sensitivity for holographic recording in different materials

Material	Spectral region nm	Light sensitivity, S^{-1} for $\eta = 1\%$ J/cm^2	Recording conditions
KCL-M	600 – 750	6.10^{-2}	$\lambda_1 = 488$ nm; $\lambda_2 = 752.5$ nm $\eta = 15\%$
KBr-F (addit. color.)	550 – 700	14	$\lambda_1 = \lambda_2 = 514.5$ nm $\eta = 12\%$
SrTiO$_3$-Ni,Mo	500	5.10^{-2}	$\lambda_1 = \lambda_2 = 632.8$ nm $\eta = 12\%$
Na$_2$O-Al$_2$O$_3$-SiO$_2$-Cl (glass)	200 – 300	5	UV spectral region
Na$_6$Al$_6$Si$_6$O$_{24}$(NaX)$_2$ X = Cl; Br; I	300 – 700	>5	The electron beam recording is also possible

M-center reorientation is one of the most light sensitive reactions used in holographic recordings ($S^{-1} = 60$ mJ/cm^2 for a 1% diffraction efficiency, Table 4.1). However, a high M-center concentration ($N_M \sim 10^{-18}$ cm^{-3}) is required. It is difficult to get such high concentrations in additively colored crystals because larger aggregates and colloidal centers are produced, depressing M-center formation. KCl and NaF crystals are the most useful specimens. However, as previously mentioned, thermal annealing and recombination of electron and hole centers disturbe optical recording in irradiated crystals.

4.1.3 Light Induced Aggregation of F-Centers

Aggregation of single F-centers to produce F_n-, X- and macroscopic metallic colloid centers are well known phenomena in alkali halide and other ionic crystals [3.13,14]. It was proposed that light induced $F \rightarrow X$ reactions may also be useful in optical data storage. X-centers are quasi-colloids which contain several hundred, or thousand, F-centers with a quasi-metallic binding (large aggregates of F-centers have a similar structure as alkali metals, Fig. 4.1[1]). These clusters are very stable, thus, are suitable for long time storage. The efficiency of the photothermal $F \rightarrow X$ reaction in additively colored alkali halide crystals is highest at 370 - 520 K. Thus, the usual processing technique is to irradiate the crystal in the F-center absorption region while heating in this temperature range. The X-centers thus created have a strong absorption band at room temperature, which can be nondestructively optically read (Fig. 4.3).

The use of an $F \rightarrow X$ reaction for an optical recording has several positive attributes [4.15]: 1) higher recording efficiencies than for F_n-centers (F_n-centers are small aggregates with $n \leq 5$); 2) nondestructive long-time information storage due to the stability of X-centers; 3) possibility of thermal and optical erasure; 4) nondestructive two- or multichannel reading (using different wavelengths in the X- and F-bands); 5) possibility of recording of phase holograms (using different wavelengths for recording λ_1 and reconstruction λ_2).

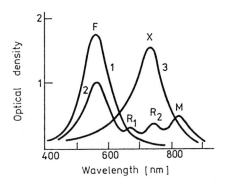

Fig. 4.3. Absorption spectra of additively colored KCl crystals: (*1*) F-center absorption before bleaching; (*2*) after illumination in the F-band at 300 K (R_1-, R_2- (clusters containing 3-F-centers) and M-centers formation); (*3*) after illumination in the F-band at 470 K (X-center formation)

---

[1] X-centers were first observed by *Scott* and *Shatalov* [4.13,14]. The optical and thermodynamical properties of theses centers are different from those of simple point defects (like F- and M-centers); their properties are common to large (macroscopic) alkali metal colloids. However, the size of X-centers is smaller and no direct size measurements for these centers are possible.

4.1.4 Phase Hologram Recording in Photochromic Materials

As mentioned in Chap. 2, every optical recording changes the complex refractive index $\tilde{n} = n - i\kappa$. Every photochromic reaction which changes the absorption index to $\Delta\kappa$, also contains a phase recording component: Δn. The relation between the dispersion[2], $\Delta n_d(\omega)$, and absorption coefficients, $\kappa(\omega)$, (Chap. 2) for a light absorbing medium, is given by the Kramers-Kronig relation [4.16]:

$$n(\omega_i) = \frac{2}{\pi} P \int_0^\infty \frac{\kappa(\omega)}{\omega^2 - \omega_i^2} d\omega, \tag{4.2}$$

where ω_i is the angular frequency of the light and P the real part of the integral. For a definite absorption band, $\Delta n_d(\omega_i)$ is determined from the experimentally measured absorption coefficient, $\kappa(\omega_i)$, at different frequencies. Then the light induced optical constant changes, Δn, $\Delta\kappa$ are simply the difference of the initial and final states of $n_d(\omega_i)$ and $\kappa(\omega_i)$, i.e., before and after the photoinduced reaction. The efficiency of the optical recording process can, of course, also be determined as a function of temperature.

A detailed analysis of the processes occurring in phase and amplitude hologram recording in additively colored KBr-F crystals for the F $\Leftrightarrow$ X reaction was made by *Ozols* [4.17]. From the $\Delta n(\lambda)$ dependence, he found that the best conditions for phase recording, which are, of course, functions of the recording and reconstruction wavelengths, the initial optical thickness and the light induced absorption change, are in the long wavelength region of the F-absorption band (Fig. 4.4a). The optical density, $D(\lambda)$, is proportional to $\kappa(\lambda)$. However, in the long wavelength region of the F-center absorption band, there is an overlap with the X-center absorption. Therefore, the short wavelength region of the F-center absorption band is preferable. Figure 4.4b shows the calculated light induced refractive index change, $\Delta n(\lambda)$, diffraction efficiency, $\eta(\lambda)$ and phase recording ratio, $\beta(\lambda) = \eta_{ph}(\lambda)/\eta(\lambda)$ (where η_{ph} is the phase part of the total diffraction efficiency) [4.17, 18]. The diffraction efficiency is much higher in the shorter wavelength region (and in the longer wavelength region) than in the maximum of the F-band. The oscillations in $\eta(\lambda)$ and $\beta(\lambda)$ are typical for volume holograms and are due to interference effects (Table 2.5).

The experimentally measured $\Delta n(\lambda)$, $\eta(\lambda)$ and $\beta(\lambda)$ in the spectral region 457.9–528.7 nm (Ar+-laser) and 632.8 nm (He-Ne laser) are in good

---

[2] The dispersion is determinated by $\Delta n_d(\omega) = n(\omega) - n(0)$, where $n(0)$ is the refractive index far from the maximum of the absorption band.

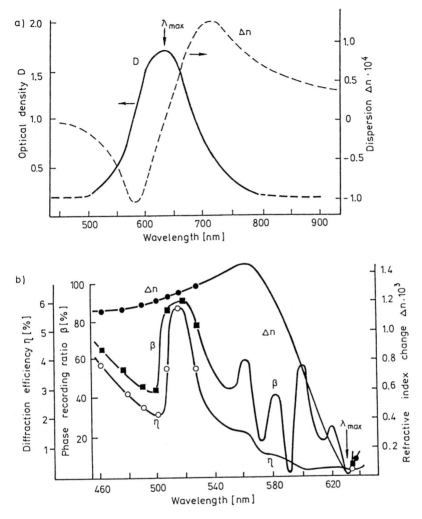

Fig. 4.4a, b. Theoretical and experimental measurements of optical   phase recording parameters in additively colored KBr-F crystals (F-center concentration: $N_F = 0.3 \cdot 10^{17}$ cm^{-3}, T = 293 K, thickness d = 3 mm). (**a**) Theoretical dependence of the optical density $D(\lambda)$ and dispersion: $\Delta n(\lambda) = n(\lambda) - n(0)$ (where n(0) is the refractive index at a wavelength far from the absorption maximum: $\lambda_{max} = 630$ nm). (**b**) Calculated spectral dependence $\Delta n(\lambda)$, $\eta(\lambda)$ and $\beta(\lambda) = \eta_{ph}(\lambda)/\eta(\lambda)$ for a recorded holographic grating with period $\Lambda = 1.5$ μm (recording at T = 450 K with $\lambda_1 = 514.5$ nm; reconstruction at 293 K at a different wavelength λ_2); experimental points for the wavelength region 457.9–528.9 nm and 632.8 nm [4.17]

agreement with calculations of these quantities. To understand the origin of theoscillations above 540 nm, further detailed experimental investigation is necessary.

The results from [4.17] demonstrate the possibility of making phase recordings in photochromic materials. Our estimation for the maximum value of the diffraction efficiency, η_{max}, in a phase hologram using the $F \Leftrightarrow X$ reaction in alkali halide crystals, is ~ 60% [4.15]. In additively colored KBr-F crystals an experimental value of $\eta = 20 - 40\%$ was obtained.

Phase hologram recording was also demonstrated to be possible in KCl crystals with M-center reorientation: by recording with $\lambda_1 = 488$ nm and reconstructing with $\lambda_2 = 752.5$ nm in KCl-M crystals of thickness $d = 0.4$ nm, a diffraction efficiency of $\eta = 15\%$ was achieved (Table 4.1) [4.18]. This technique of recording and reconstruction with different wavelengths is widely used in real time holography.

4.2 Photorefractive Materials

The observation of photorefraction in ferroelectric crystals, that is, when upon illumination at $\Delta n >> \Delta \kappa$, spurred the development of a new class of phase recording materials. These ferroelectric materials have a high polarizability and a single domain structure. Most of them are oxides (Table 4.2). For optical recording, a stable ferroelectric phase at room temperature (the Curie point must be higher than 400 K) is desirable as well as high transmission in the visible region. However, high transmission implies low light sensitivity. Therefore, impurity doping is used to optimize the recording material. Photorefractive ferroelectric materials can be used for phase hologram recording as well as for dynamic holography [4.19]. Below, some general properties and the photorefraction mechanisms of these materials are discussed.

X-ray scattering analysis has yielded a simple structural model for oxide ferroelectrics. From this model it has been found that the main structural element is the BO_6 octahedron (Fig. 4.5a), where B is a transition metal atom. For example, perovskites (Table 4.2) have the empirical formula $ABO_3 = AO \cdot BO_2$, where the cations (A) form a cage about a BO_6 octahedral unit. For $LiNbO_3$, the structural unit is $Li_2O \cdot Nb_2O_5$, for $BaTiO_3$, it is $BaO \cdot TiO_2$. The same BO_6 model can also be used for tungsten bronze type compounds, an example being, $SrNb_2O_6$ can be described as an AB_2O_6 crystal. $Bi_{12}GeO_{20}$ and $Bi_{12}SiO_{20}$ crystals have a tetrahedral structural element, MO_4, where M is the Si or Ge atom [4.20].

The BO_6 octahedra or MO_4 tetrahedra determine the main properties of the crystal. Defect states which allow optical recording are also connected with these structure elements. As an example, in $LiNbO_3$, the Nb atoms in BO_6 can be displaced by impurity atoms (e.g., Fe, Cu) or the oxygen atom by a

Table 4.2. Photorefractive ferroelectric materials as (the Curie point T_C in K is given in parentheses)

Non-Oxides Antimony sulphur iodides	Perovskites	Oxides Lithium-niobates	Complex oxides with tungsten bronze structure	Bismuth germanium compounds	PLZT (PbLaZrTi) ceramics
SbSI (293)	$BaTiO_3$ (408)	$LiNbO_3$ (1503)	$Sr_xBa_{1-x}Nb_2O_6$	$Bi_{12}GeO_{20}$	PLZT:La[a]
SbSeI (296)	$CaTiO_3$	$LiTaO_3$ (938)	[SBN; for x=0.75 (395)]	$Bi_{12}SiO_{20}$	
BiSI (113)	$KNbO_3$ (708)				
	$KTa_xNb_{1-x}O_3$ [KTN; for x=0.5 (360)]				

[a]T_C depends on the composition; optical recording in PLZT occurs in an external field

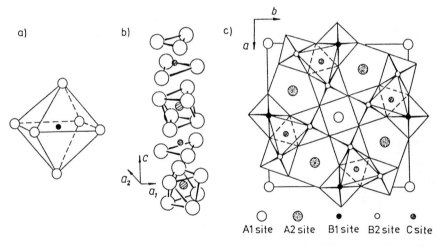

Fig. 4.5a–c. The ferroelectric oxide crystal structure. (**a**) The BO_6 structural unit for ferroelectric oxides ABO_3, AB_2O_6 and $A_6B_{10}O_{30}$, where (A) are metal atoms (not pictured) and B are transition metal atoms shown by the *filled circle*. *Open circles* denote positions of oxygen atoms. (**b**) $LiNbO_3$ and $LiTaO_3$ crystal structure, the *small hatched circles* represent Li ions, the large *hatched circles* are Nb or Ta and the *large open circles* O ions. (**c**) The tungsten bronze structure looking down the tetragonal c-axis. Examples of species with such structures are the complex ferroelectric oxides $Ba_2Sr_3Nb_{10}O_{30}$, $NaBa_2Nb_5O_{15}$ etc. For the complex oxide $K_xLi_yNb_zO_{30}$ the cations K are in two positions (A1 and A2), Li (A1, A2 and C) and Nb in three positions (C,B1 and B2). From [4.20]

vacancy. Such defects have a strong influence on the light sensitivity and the mechanism of optical recording. In recent years, a progress in the understanding of the impurity defect microsructure in $LiNbO_3$ crystals has accumulated. Simple impurity centers of the transition elements (Fe, Mn, Co, Ni, Cr) usually substitute Li atoms in the lattice [4.21]. However, for optical recording, thermal annealed crystals (in different atmospheres) are used, which have a more complex defect structure (impurity and intrinsic defect aggregates). Thus, at present, many optical recording processes are only phenomenologically understood .

4.2.1 The Mechanisms of Photorefraction in Ferroelectrics

In photorefractive materials the change in the refractive index can be induced by several mechanisms: (1) a photovoltaic effect, that is a light induced voltage caused by an asymmetric charge transfer; (2) diffusion of charge carriers in a concentration gradient or (3) a drift of charge carriers in an external electric field [4.19]. These are the main mechanisms which have been investigated in greater detail in $LiNbO_3$ and $LiTaO_3$ crystals. Birefringence in these, and other ferroelectric crystals, can also be induced by irradiation with X-rays, fast electrons or other energetic particles. This was observed by *Ohmori* et al. [4.22] and further investigated by *Volk* et al. [4.23]. In $LiNbO_3$, a strong influence of the Fe concentration in the X-ray induced birefringence was observed. X-ray induced birefringence was also observed in other monocrystals such as $Ba_2NaNb_5O_{15}$-Fe, Mo, and triglycine ($C_6H_{11}N_3O_4$) sulphate. The phenomenological mechanism of photorefraction in $LiNbO_3$ and $LiTaO_3$ crystals was proposed by *Chen* [4.24]. The light induced birefringence results from an intrinsic electric field which is created by the charge carrier transport to defect states. This light induced electric field leads to an intrinsic electro-optic effect [4.20]. The questions of why the charge carriers separate, that is, the nature of the local field and the microstructure of the trapping centers were still unanswered. However, it was known that the light induced birefringence was an intrinsic linear electro-optic (Pockels) effect determinated by the electro-optic tensor (Sect. 3.4.1).

The spatial distribution of the light induced birefringence: $\Delta n = \Delta(n_e - n_o) = f(r)$, over an illuminated area with a diameter of $2r$ of $LiTaO_3$-Fe, is shown in Fig. 4.6. The curve, in part b, demonstrates that the light induced spatial distribution of the charge leads to different signs of Δn inside and outside the illuminated region. Thus, the intrinsic electric field inside and outside of the illuminated area is also pointed in different directions. Further investigations by *Glass, von der Linde* and *Negran* showed that in $LiNbO_3$ and $LiTaO_3$ photorefraction is caused by an anomalous bulk photovoltaic effect [4.25]. *Glass* and *von der Linde* proposed that the photovoltaic effect in Fe doped $LiNbO_3$ and $LiTaO_3$ crystals is caused by Fe atoms which displace the Nb and Ta atoms in the BO_6 octahedra to noncentrosymmetric positions. However, the

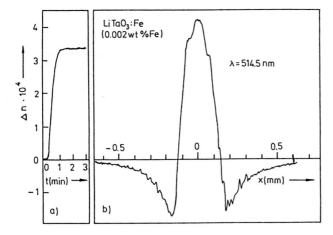

Fig. 4.6a,b. Photorefraction of LiTaO₃-Fe (0.002 weight %) crystals. (**a**) Time dependence of the birefringence change, Δn, induced by a parallel Gaussian laser beam with a diameter of 0.26 mm. (**b**) Spatial dependence of the saturation value of Δn, measured with a focused Gaussian laser beam with a diameter of 15 μm. Excitation with Ar^+ laser light at $\lambda = 514.5$ nm

details of the mechanism of the bulk photovoltaic effect in LiNbO₃ and other ferroelectric materials has not yet been investigated. Further investigations showed that photorefraction can be enhanced by an external electric field which is applied in the same direction as the intrinsic bulk photovoltaic field [4.20].

Holographic recording in LiNbO₃ and other ferroelectric crystals was proposed to proceed by a light induced carrier diffusion mechanism [4.19]. After recording an elementary holographic grating with a period Λ, the charge carrier concentration at the interference maxima is much larger than at the minima, resulting in a concentration gradient. This diffusion mechanism usually, however, is important only for small grating periods in materials where the bulk photovoltaic effect is absent. Let us discuss, in more detail, these main photorefraction mechanisms. By illumination of a crystal with $I(x)$ (x is a crystal coordinate) the induced current density is given by:

$$j(x) = \kappa_{ph\text{-}v}\kappa_a I(x) + \sigma(x)(E_{int} + E_{ext}) + eD\frac{dN}{dx},$$

(4.3)

where $k_{ph\text{-}v}$ is the photovoltaic or Glass constant; k_a the absorption coefficient; $\sigma(x)$ the conductivity; E_{int} and E_{ext} the intrinsic (internal) and external electric field strengths; e the elecron charge; D the electron (or hole) diffusion coefficient; N the concentration of charge carriers. The first term in (4.3) describes the bulk photovoltaic effect [4.20,25], while the charge carrier transport, due to drift in an internal or external electric field or diffusion due to a concentration gradient [4.19], are described by the second and third terms. The current density depends on the composition of the ferroelectric material as well as on the excitation conditions such as the light intensity and pulse length, the temperature, and the direction of illumination with respect to the optic axis.

The terms in (4.3) can be time dependent, e.g., the intrinsic field E_{int} in a material increases with the illumination time and a correlation between Δn and E_{int} takes place. We will now review experimental studies of these mechanisms in more detail.

Photovoltaic Mechanism. In ferroelectrics the anomalous photovoltaic effect induces a very high potential (several thousans volts and more) and is quite different from that seen in semicondutors. This is because the low conductivity and high breakdown voltage forces the voltage gradient from the localized charge carriers. The anomalous photovoltaic effect was first observed by *Chynoweth* and *Fridkin* in $BaTiO_3$ and $SbI_{0.35}Br_{0.65}$ crystals [4.20]. The field arises from the nonsymmetrical crystal structure at defect sites. The correlation between photoconducting properties and the photoinduced change in the refraction index in Fe and Cu doped $LiNbO_3$ crystals was observed experimentally by *Glass* et al. [4.25]. They noted a constant stationary photocurrent in doped short-circuit $LiNbO_3$ crystals ($j \sim 10^{-9}$ A/cm^2) under illumination with 514.5 nm Ar$^+$-laser light. Open circuit crystal illumination leads to both a photovoltage and photorefraction.

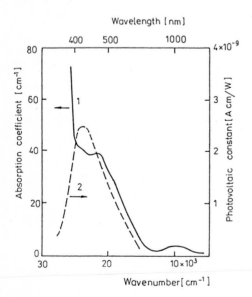

Fig. 4.7. The absorption spectrum of iron doped $LiNbO_3$ crystal (*1*) and the photovoltaic constant κ_{ph-v} (*2*). From [4.20]

The change in the birefringence, induced by light, in ferroelectrics is employed in optical recordings. This is a complex process in the anomalous photovoltaic effect, which is fixed by the defect microstructure and the anisotropy of the host crystal. The defect microstructure determinates the spectral light sensitivity, i.e., the spectral dependence of the photovoltaic constant. The induced intrinsic field, due to the birefringence via the electro-optic effect, which is determinated by the electro-optic tensor of the host crystal (Sect. 3.4.1). Therefore, both the local defect symmetry and the anisotropy of the crystal influence the photorefraction. The spectral dependence of the photovoltaic effect in doped $LiNbO_3$ and $LiTaO_3$ was measured by *Glass et al.* [4.20]. The maximum value of κ_{ph-v} corresponds to the maximum of the absorption spectrum of the photovoltaic defects (probably Fe^{2+} or their complexes); a shift of $\kappa_{ph-v}(\lambda)$ with respect to the defect absorption spectrum is caused by the fundamental absorption band.

A detailed analyses of the photovoltaic effect and optical recording in doped $LiNbO_3$ crystals is given by *Festl, Hertel, Krätzig* and *von Baltz* [4.26]. It was demonstrated, that the photorefraction effect (i.e., the intrinsic electro-optical changes of Δn) is determined only by the host crystal anisotropy, while the spectral dependence and the photovoltaic field are determined by the local defect symmetry. Such a general concept can be used for all ferroelectric crystals.

It is interesting to estimate the intrinsic photovoltaic field, E_{int}, in the crystals. The magnitude of E_{int} is given by the classical equation of electrodynamics [4.29]:

$$E_{int} = \left(\varepsilon\varepsilon_0^{-1}\right)\int \rho_e(x)dx = \left(\varepsilon\varepsilon_0^{-1}\right)\int j(x)\,dt, \qquad (4.4)$$

where $\rho_e(x)$ is the charge density, ε, ε_0 the dielectric constant and the permittivity of vacuum, x is the coordinate in the crystal (often the direction of x is chosen to be parallel to the optical axis, $x \| c$), and t is the time. If the photovoltaic effect dominates and $E_{ext} = 0$, we get from (4.3):

$$j(x) = k_{ph-v}\kappa_a I + \sigma E_{int}. \qquad (4.5)$$

The light induced electric field, E_{int}, reduces the current, giving: $j(x) \to 0$ for saturation. Thus, we get from (4.5) the magnitude of E_{int}:

$$E_{int} = \frac{k_{ph-v} \cdot \kappa_a I}{\sigma}, \qquad (4.6)$$

where the conductivity term is:

$$\sigma = \sigma_d + \sigma_{ph} = \sigma_d + \beta I, \tag{4.7}$$

σ_d, σ_{ph} are the dark- and the photoconductivity and β is the quantum efficiency of photoconductivity. For a crystal with a low dark conductivity ($\beta I \gg \sigma_d$), we can estimate, from (4.6), the magnitude of E_{int} as:

$$E_{int} = \frac{k_{ph-v} \cdot \kappa}{\beta}. \tag{4.8}$$

Therefore, at high incident light intensities, the intrinsic photovoltaic field does not depend on the light intensity. This was experimentally demonstrated for LiNbO$_3$ and LiTaO$_3$ crystals [4.27 – 30].

The photorefraction kinetics in LiNbO$_3$ crystals depends on the light intensity, time of exposure and temperature [4.28]. Δn increases with light intensity and reaches a larger magnitude of saturation, Δn_{sat} (Fig 4.8).

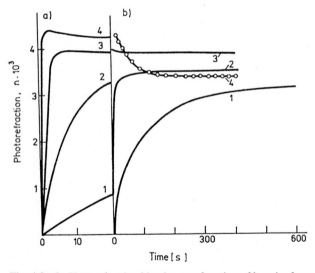

Fig. 4.8.a,b. Photorefraction kinetics as a function of length of exposure and light intensity at $\lambda = 514.5$ nm in LiNbO$_3$-Fe (0.1 weight %): (**a**) initial stage; (**b**) saturation stage. Curves (*1*) 0.3 W/cm^2; (2) 3 W/cm^2; (3) 300 W/cm^2; (4) 10^4 W/cm^2

However, at high intensities ($I > 10$ W/cm^2) the saturation value, Δn_{sat}, is smaller than the maximum value, Δn_{max}, at smaller exposure times. The inequality: $\Delta n_{max} > \Delta n_{sat}$, leads to light induced heating which anneals the initial photorefraction. The slope: $\Delta n = f(I, t)$, is steeper for small exposure times; in the saturation region, Δn_{sat} reaches, approximately, the same

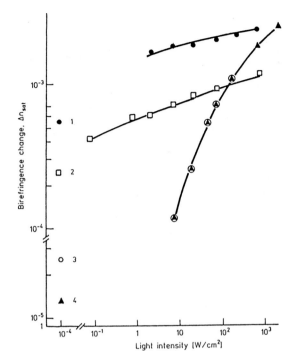

Fig. 4.9. Saturation values of Δn_{sat} as a function of light intensity (at $\lambda = 514.5$ nm) for various LiTaO$_3$-Fe crystals: (1) 0.002 weight % Fe, $k = [Fe^{2+} / Fe^{3+}] = 1$; (2) 0.002 weight % Fe and $k \approx 0.01$; (3) 0.02 weight % Fe and $k \approx 0.2$; (4) 0.05 weight % Fe and $k \approx 0.2$

value $(3-4)\cdot10^{-3}$ for all light intensities. Similar behavior is observed also in LiTaO$_3$-Fe crystals (Fig. 4.9); here, Δn_{sat} depends stronger on the light intensity than in LiNbO$_3$-Fe crystals. However, at light intensities of 10^4 W/cm^2 they are of the same magnitude ($\Delta n_{sat} = (1 - 3)\cdot10^{-3}$) as in LiNbO$_3$ crystals. We can correlate Δn_{sat} to the intrinsic electric field, with:

$$E_{int} = \alpha \cdot \Delta n_{sat}, \tag{4.9}$$

where α is a coefficient depending on the direction of propagation of light in the crystal and the direction of polarised light. For light polarized at a 45° angle to the optical axis, c, propagating in the direction perpendicular to c, we found the relation of α to the electro-optical properties in the crystal ([3.20] p. 282):

$$\alpha = \frac{2}{n_e^3 r_{33} - n_o^3 r_{13}}, \tag{4.10}$$

where n_e, n_o are the extraordinary and ordinary refractive indices and r_{13}, r_{33} the electro-optic coefficients (Table 3.9). The electro-optic coefficients, r_{ij}, and the coefficient α are practically independent of the light intensity (up to 10^4 W/cm^2). For LiTaO$_3$, the experiment gives $\alpha = 6.3 \cdot 10^7$ V/cm ($\pm 10\%$) and is practically constant in the spectral region 450 - 630 nm. For LiNbO$_3$ crystals, $\alpha = 9 \cdot 10^7$ V/cm at $\lambda = 633$ nm and weakly decreases at shorter wavelengths (at $\lambda = 500$ nm the decrease is approximately 20%) [4.28]. These experimental results are in a good agreement with (4.10).

Our estimation of the light induced intrinsic electric field from the experimentally measured Δn_{sat} and Δn_{max} for LiTaO$_3$ and LiNbO$_3$ crystals (according to (4.9) in both cases: $\Delta n_{sat} \approx 3 \cdot 10^{-3}$) coincides with the photovoltaic current measurements (Glass et al. [4.25]) and gives values of 200-270 kV/cm, which are the largest experimentally observed electric fields in ferroelectric crystals [4.20, 28].

The light induced electric fields, E_{sat}, in these crystals are several orders of magnitude smaller than their electric breakdown fields. However, electrical discharges in the atmosphere surrounding the surface of the illuminated LiNbO$_3$-Fe crystal have been observed (the breakdown voltage of air is $\approx$ 30 kV/cm which is much smaller than E_{sat}) [4.30]. Such discharges puts a limit on the electric field strength used in phase recording (an electric breakdown will lead to information losses).

Diffusion Mechanism. When the photovoltaic effect is absent and there is no external field, charge carrier diffusion becomes the dominant mechanism of photorefraction in ferroelectric materials. According to (4.3), the diffusion current density, $j(x)$, is:

$$j(x) = eD \frac{dN}{dx}. \tag{4.11}$$

The charge carrier concentration, $N(x)$, is proportional to $I(x)$. For an elemental holographic grating $I(x) \sim \cos(2\pi x/\Lambda)$ (Sect. 2.4.2). According to (4.11), the final light induced diffusion field, E_D, and refractive index phase grating $\Delta n(x) \sim dN/dx$ is inversely proportional to the grating period $1/\Lambda$ and has a $\pi/2$ phase shift with respect to the light interference field $I(x)$. This has been verified in many investigations [4.19, 24, 29]. In LiNbO$_3$ and LiTaO$_3$ crystals, where both mechanisms (photovoltaic and diffusion) are possible, the diffusion mechanism dominates at grating periods of $\Lambda \leq 1\mu$m (Table 4.3).

Recording in an External Electric Field. An external electric field can enhance the bulk photovoltaic effect or it can induce a charge carrier drift [4.19, 31, 39]. In PLZT ceramics, for example, the drift mechanism is usually the dominant process by which holographic information is stored [4.36, 40].

Table 4.3. Photorefraction mechanisms in ferroelectric materials

Mechanism	Minimal energy density for photoinduced $\Delta n = 10^{-5}$ [J/cm^3]	Material	Recording conditions
Bulk photovoltaic effect	0.290	LiNbO$_3$-Fe [4.31]	Impurity center or band-to-band excitation by holographic recording with gratings periods $\Lambda \geq 1\ \mu m$
Diffusion of charge effect	0.005	LiTaO$_3$-Fe [4.32] KNbO$_3$-Fe [4.34]	Holographic recording on short circuited crystals at periods $\Lambda < 1\ \mu m$
Drift of charge carriers in an external electric field	$0.76 \cdot 10^{-3}$	LiNbO$_3$ [4.35] LiNbO$_3$ [4.35]	Holographic recording in an electric field at periods $\Lambda < 1\ \mu m$
	12	PLZT ceramics (PbLaZrTi) [4.36]	Holographic recording in an electric field at periods $\Lambda \approx 0.5 - 4\ \mu m$
Photoinduced pyroelectric field	–	LiNbO$_3$-Fe [4.37] LiNbO$_3$ [4.38]	Holographic recording at large periods $\Lambda > 100\ \mu m$
Dipole moment changes	1	LiNbO$_3$-Cr^{3+} [4.31, 33] LiTaO$_3$-Cu^{2+} [4.31, 33] KNbO$_3$-Fe^{3+} [4.31, 33]	Pulsed laser excitation

Light Induced Dipole Moment Changes. In addition to longer-range electron-hole generation and charge carrier diffusion, the local optical excitation of polar point defects can induce a polarization change which leads to a change in the refractive index [4.42]. Such a local change in the refractive index is modulated by the anisotropy of the crystal (i.e., the polarizability of the medium determined by the index ellipsoid, Fig. 3.10). The light induced dipole excitation mechanism was observed experimentally in LiNbO$_3$ crystals under pulsed laser excitation [4.32].

The main photorefraction mechanisms in some ferroelectric materials are presented in Table 4.3. Note that in a given material under different recording conditions (e.g., light intensity and intensity distribution, or presence of an external field) various recording mechanisms are possible. A general property of the photorefraction phenomenon in ferroelectric crystals is that the light induced birefringence (i.e., the intrinsic electro-optic effect) is determined by the electro-optic tensor of the crystal (Chap. 3). The local changes of the defect states (electric field, symmetry) are much smaller than the anisotropic properties of the host crystal.

4.2.2 Optical Recording in LiNbO$_3$

Early investigations into photorefraction processes in LiNbO$_3$ revealed the importance of impurity centres [3.19, 20]. Different impurities (e.g., Fe, Cu) which have absorption peaks near the band gap influence the light sensitivity of the crystals. In Fig. 4.10, the dichroism of LiNbO$_3$ and LiNbO$_3$-Fe crystals is demonstrated. For optical phase recording, small impurity concentrations

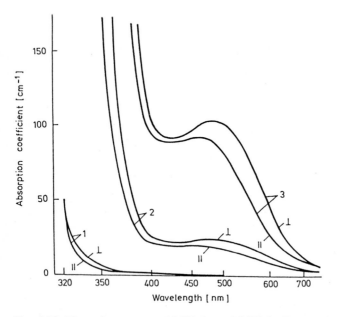

Fig. 4.10 Absorption spectra of LiNbO$_3$ and LiNbO$_3$-Fe crystals: (*1*) undoped; (*2*) 0.1 percentage weight Fe; (*3*) 0.3 percentage weight Fe; ||, ⊥ - direction of light vector with respect to the optic axis

are used and excitation occurs in the visible region ($\lambda \geq 400$ nm) [4.19, 42]. In LiNbO$_3$-Fe crystals, the impurity centers act as donors and acceptors (most likely Fe^{2+} are donors and Fe^{3+}-centers act as the electron acceptors). The ratio of Fe^{2+}/Fe^{3+} depends on the thermal treatment of the crystal (annealing in oxygen or inert atmospheres). Thus, a holographic recording can be made by irradiating the Fe^{2+}-centers in LiNbO$_3$-Fe crystals with polarized light, i.e., using the optical anisotropy of the crystal. The largest diffraction efficiency can be reached by recording with polarized light perpendicular to the optic axis ($E \perp c$) and reconstructing, or reading, with polarized light parallel to c ($E \| c$) (Fig. 4.11). It is difficult to find the optimum concentration of Fe impurities. Some investigations reveal an indirect influence of the impurities on the

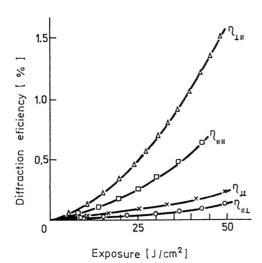

Fig. 4.11. Dependence of the efficiency of diffraction on light polarization in $LiNbO_3$-Fe (0.05 percentage mass) crystals. The indices (||) and ($\perp$) give the light vector direction with respect to the optic axis during recording (first index) and reconstruction (second index)

photorefraction efficiency due to changes in the vacancy and interstitial defect concentration. However, the maximum light sensitivity in $LiNbO_3$-Fe crystals at a percentage mass of Fe ions of 0.15, with Fe^{2+}/Fe^{+3} ratio from 0.25 to 0.40. At higher Fe^{2+} concentrations, the recording is not stable due to the large increase of dark conductivity. For example, at an Fe^{2+} concentration of 0.2 percentage mass, the storage time in $LiNbO_3$-Fe is approximately 1000 hours, while at 0.3 percentage mass, it is not more than seconds [4.38, 42].

4.2.3 Recording of Phase Holograms

Holographic phase recording in ferroelectric materials is of particular interest, given by their high diffraction efficiency, excellent resolution and long storage time (Table 4.4). In these crystalline materials, three dimensional (3-D) holograms can be recorded and new holographic memory systems can be developed. However, many crystalline materials are too expensive for practical applications due to complicated growing techniques, and three dimensional optical memory systems (Chap. 2) are more complicated than 2-D systems. Many ferroelectric materials have a high refractive index which leads to high reflection losses (for $LiNbO_3$ and $LiTaO_3$ approximately 17% from each surface). This limits the application of these materials without antireflection coatings. These drawbacks can be made less serious, as we discuss below.

The light sensitivity can usually be increased by impurity doping. However, this leads to an increase of the absorption coefficient and, hence, to a decrease in the diffraction efficiency. To regain the maximum diffraction efficiency, the

Table 4.4. Holographic recording parameters in ferroelectric materials at room temperature

Material	Light sensitivity S^{-1} for $\eta=1\%$ J/cm^2	External field kV/cm	Recording wavelength nm	Storage time	Maximum refractive index change n_{max}	Reference
LiTaO$_3$-Fe	$1.1 \cdot 10^{-2}$	15	351	10 years	10^{-3}	[4.43]
KTa$_{0.65}$Nb$_{0.35}$O$_3$ (KTN) (cubic)	10^{-4}	10	400	1 year	–	[4.44]
KTN (tetragonal)	$2 \cdot 10^{-4}$	10	400	0.6 years	–	[4.44]
BaTiO$_3$	$5 \cdot 10^{-2}$	10	458	15 hours	$2 \cdot 10^{-5}$	[4.42]
Sr$_x$Ba$_{1-x}$Nb$_2$O$_6$-Ce	$1.5 \cdot 10^{-3}$	0	488	0.1 year	10^{-5}	[4.45]
Bi$_{12}$SiO$_{20}$	$3 \cdot 10^{-4}$	6	514	1 day	–	[4.46]
K(D$_{0.7}$H$_{0.3}$)PO$_4$ (KDP)	$5 \cdot 10^{-3}$	–	300	7 days[a]	–	[4.47]
PLZT ceramics	0.1–0.6	10	400-700	Years	10^{-3}	[4.19]

[a] At 113 K

reconstruction wavelength λ_2 is made different from the recording wavelength λ_1 ($\lambda_2 > \lambda_1$, Chap. 2). However, this is possible only for 2-D holographic gratings. Another method to improve the light sensitivity is to place the medium in an external electric field while recording. In KTN (KTa$_{0.65}$Nb$_{0.35}$O$_3$) crystals, a sensitivity of $S^{-1} = 10^{-4}$ J/cm^2 (for $\eta = 1\%$) was obtained with this method [4.44], which is close to the theoretical limit for ferroelectric materials (Table 4.4) [4.48].

The dynamic range in 3-D phase holograms is defined as the value of Δn_{sat} (see (4.9) and Figs. 4.8,9). It determines both the diffraction efficiency and the number of holograms that can be recorded in a given volume of the medium. Volume phase recordings can be made with a diffraction efficiency close to 100% (however, reflection losses from the surface must be taken into account). For 3-D phase holograms, (Table 2.5) the condition for maximum diffraction efficiency of the first maximum (m = 1) is:

$$n_A = \frac{\lambda}{2d},\tag{4.12}$$

where n_A is the amplitude of the recorded phase grating and d is the thickness of the recording medium. For a recording medium with $d = 1$ mm at $\lambda = 632.8$ nm, from (4.12), we have: $n_A \approx 3 \cdot 10^{-4}$ which is one order of magnitude smaller than the saturation value of $\Delta n_{sat} = 3 \cdot 10^{-3}$ in LiNbO$_3$ and LiTaO$_3$ crystals.

By multihologram recording, with a diffraction efficiency for each hologram of $\eta_0 = 0.1\%$ in a medium with the maximum diffraction efficiency $\eta_{max} = 100\%$, 1000 holograms can be recorded. Therefore, in the dynamic range $\Delta n = 3 \cdot 10^{-3}$, each phase hologram has an amplitude $n_{Ai} = \Delta n/1000 \approx 3 \cdot 10^{-6}$. These results are close to those of *Krätzig* [4.42, 43]. In perfect ferroelectric $LiNbO_3$ crystals, with a thickness of 0.5 cm, more than 500 holograms at different angles of incidence with a signal-to-noise ratio of 40 dB, could be recorded.

The storage time depends on the types of defect in the recording material. For long-time stroage (years), one must have deep capture levels and no shallow traps for carriers of the opposite charge (for example, shallow hole traps can induce information losses from deep electron traps). Usually, there is a background of shallow traps due to crystal impurities, hence, long-time storage is possible only in very pure crystals. This was demonstrated experimentally by *Krätzig* et al. for KTN crystals. In high qualitiy ferroelectric crystals, the storage time is of the order of one year, which is acceptable for many applications, but not enough for archival memory systems. For such applications only the best $LiTaO_3$-Fe crystals, with a storage time of 10 years, can be used.

The capability of nondestructive reading is one of the most important parameters for holographic memory systems. However, for most real-time recording materials, postrecording light sensitivity leads to information losses during the readout [4.19]. There are several methods available for the nondestructive readout of ferroelectrics. One of these involves the use of different wavelengths for recording and readout. As previously mentioned, usually, $\lambda_2 > \lambda_1$, and at the readout wavelength, the postrecording light sensitivity is zero. However, this can be done only for 2-D holograms; for 3-D holograms the dispersion region $\Delta\lambda$ is very small and readout is possible only at the same wavelength (Sect. 2.4.4). *Amodei* and *Staebler* developed a thermal fixing method for $LiNbO_3$, $Ba_2NaNb_5O_{15}$ and other crystals [4.49,50]. The fixing process includes two steps: 1) thermal heating at $100°$ C for several minutes; 2) subsequent illumination with a uniform light beam. This leads to stable atomic and electronic structure, i.e., to an equilibrium polarization and a non-destructive hologram. However, in practice such thermal treatment is often not possible. For this reason, electric fixation, in particular, $Sr_xBa_{1-x}Nb_2O_6$ crystals, proposed by *Thaxter*, is a more common method [4.39].

Another nondestructive optical recording method proposed by *Krätzig* uses two-photon excitation [4.51]. This method was tested in weakly doped $LiNbO_3$-Fe and $LiTaO_3$-Fe crystals for long-time information storage (at low Fe concentration the dark conductivity is small and, therefore, the storage time is long). Two-photon excitation is used to excite the electrons of an Fe^{2+}-impurity into the conduction band. Large incident light intensities are necessary (10^8 W/cm^2) to increase the probability of two photon excitation of Fe. Unfortunately, LiNbO$_3$-Fe crystals possess simultaneously induced additional color centers, which lead to an increase in dark conductivity with a subsequent loss of information (the storage time is only 10 hours). LiTaO$_3$-Fe (0.002 percentage weight) crystals do not have such disturbing effects and two-photon excitation allows long-time optical storage. In these crystals, an elementary holographic grating ($\Lambda = 10$ μm) was produced with the first harmonic of an Nd: YAG laser pulse ($\lambda_1 = 1.06$ μm, pulse length $\tau = 30$ ns; 30 mJ/cm^2). The second photon was simultaneously produced by the second harmonic of the Nd:YAG laser ($\lambda_2 = 530$ nm, pulse length 30 ns; 100 mJ/cm^2) to excite the ^{5}D electronic level of the Fe^{2+} ions. This double excitation generated a stable holographic recording with ionized Fe^{2+}-centers and electrons captured in deep traps. A light sensitivity of $S^{-1} \approx 3$-6 J/cm^2 for $\eta = 1\%$ was obtained. The two-photon recording method can be used for many recording materials. However, this method is much more expensive than one-photon recording (an expensive laser setup is necessary) and the wavelength of the second harmonics must be in resonance with the defect states (two-photon absorption spectrum) in the recording medium. At present, this method is not used for practical applications.

Reversible Information Recording. In all ferroelectric recording materials, repeated recordings are possible after thermal erasing. Usually, the medium is heated at 570-720 K for several minutes [4.19]. In many applications, however, such high temperature erasing is not possible. Therefore, optical erasing is more suitable. *Huignard* et al. [4.52] have proposed a particularly interesting erasure process in which an initial recorded hologram is overwritten by a second recording of the same hologram shifted by a half wave. Then, the interference maximum of the first hologram is at the minimum of the second hologram. This method is particularly interesting for real-time holographic memory systems.

4.3 Chalcogenide Semiconductors

Semiconductor materials are used for real-time optical recording, as well as for optical processing in multilayer systems and switching devices [4.53, 54]. Special interest is evoked by chalcogenide semiconductors, used for optical recording and also as photoresists. Before discussing these species in detail,

Table 4.5. Photoinduced phase transitions in semiconductor films for optical imformation recording

Process	Material	Sensitivity S^{-1} J/cm^2	Reference
(1) Amorphous state $\Leftrightarrow$ Crystalline state (a $\Leftrightarrow$ b)	Sb_xSe_{1-x}	0.01-1.00	[4.55]
(2) Amorphous state 1 $\Leftrightarrow$ Amorphous state 2 (a1 $\Leftrightarrow$ a2)	As_xSe_{1-x}	0.01-1.00	[4.56]
(3) Amorphous state 1 $\rightarrow$ Amorphous state 2^a (a1-M $\rightarrow$ a2-M)	As_xSe_{1-x}	0.001-0.1	[4.57]
(4) Crystalline state 1 $\Leftrightarrow$ Crystalline state 2 (c1 $\Leftrightarrow$ c2)	Sb-In	0.025	[4.58]
(5) Solid state $\rightarrow$ Gaseous state (s $\rightarrow$ v)	Te	0.13	[4.59]
(6) Metallic state $\Leftrightarrow$ Semiconductor state (m $\Leftrightarrow$ s)	SmS	0.3	[4.60]

^{a} Chalcogenide film on a metal substrate

we direct the reader to Table 4.5 which summarizes typical reactions used for optical recording in semiconductor films. Reactions (1), (4) and (5) are used for producing optical memory discs (the non-reversible reaction (5) only for production of master discs). These reactions are of photothermal origin (light absorption $\rightarrow$ phonon generation (optical heating) $\rightarrow$ local phase transition) and will be discussed in Chap. 5.

Reaction (2) is typical for many amorphous chalcogenides. The light sensitivity of these reactions can be increased when a semiconductor film is placed on a metal substrate or on a metallic film (reaction (3)). The increase of light sensitivity by approximately one order of magnitude is related to the presence of an additional metallic layer on the chalcogenide semiconductor. However, such metal atom diffusion reactions are non-reversible [4.57].

4.3.1 Atomic Structure

The amorphous chalcogenide semiconductors were discovered by *Kolomiets* and *Goryunova* in the 1950s [4.61, 62]. This was the first class of disordered semiconductors to be found. They can be produced with widely varying stoichiometry in binary A_xB_{100-x} and more complex compounds. These materials have a layered structure (similar to glassy selenium) and are also called Se-like systems[3] [4.62]. Chalcogenide semiconductors can be produced by cooling molten solids (glasses), vacuum deposition (thin films), chemical reactions (precipitation from liquid solutions). The atomic structure is different in bulk disordered systems and in thin films. Usually, the degree of disorder, determined by the medium range structure ($0.5 \leq r \leq 2.0$ nm), is higher in thin films than in bulk materials. In both cases, the physico-chemical properties of amorphous materials are determined primarily by the short range order, $r \leq 0.5$ nm.

The atomic structure of amorphous solids is characterized by the same parameters which are used for crystalline materials: coordination number N_c, valency, bond length ($r_{A\text{-}B}$ or coordination radius r_c), bond angle (θ_A, θ_B). However, the atomic structure of amorphous solids is different for each atom i, and the whole structure of the disordered material is a superposition of local structural elements (N_c, $r_{A\text{-}B}$, θ_A). The coordination radius, r_c, and valence angles, θ, for different atoms in the amorphous state vary approximately by $\pm 5\%$ for $\Delta r_c/r_c$, and $\pm 10\%$ for $\Delta\theta/\theta$. The valency of atoms and stoichiometry (for binary or more complex compounds) also vary in the amorphous state. Such variable structure units are the topological elements in the isotropic random network structure of amorphous solids.

Amorphous materials do not have a well-defined melting point. The phase transition from the solid to the liquid state takes place gradually at a temperature of $T > T_g$, the temperature of the transition of glass. These materials are sometimes referred to as "supercooled liquids" [4.64,65]. Like supercooled liquids, amorphous solids are thermodynamically unstable systems, which have a tendency to relax to the crystalline state. However, the relaxation time for many amorphous solids is extremely long – thousands of years for oxide glasses and much longer for amorphous minerals such as, for example, obsidian. Therefore, many of these noncrystalline solids are stable, for all practical purposes.

The light sensitivity of as-evaporated chalcogenide films (As_2S_3, As_2Se_3, etc.) is approximately one order of magnitude higher than for annealed films. This can also be explained by the metastable structure of the as-evaporated films which leads to larger changes in the optical properties, upon illumination.

---

[3] Si-like amorphous semiconductors (Si, Ge, SiH, SiC etc.) which have a random network structure with quadruply coordinated Si or Ge atoms will not be discussed here. These semiconductor materials have an extremely large tendency to crystallize. They cannot be formed by quenching from the melting phase because crystallization intervenes before a glass transition takes place. a-SiH semiconductors which are used in solar cells and other devices are more stable in the amorphous state.

The main problem associated with the physics of amorphous materials is to find the correlation between the atomic structure (short and medium range order) and the physico-chemical properties, which we discuss below. Let us start with a short analysis of the atomic structure of crystalline and amorphous chalcogenides.

Elemental Chalcogens. At standard conditions, all chalcogen elements, excluding oxygen, are solids with a pronounced allotropy. The valency, N_υ, of chalcogen atoms S, Se and Te is equal to 2. For sulphur and selenium, the semiconductor properties are dominant, whereas, tellurium has more pronounced metallic properties. In compounds with metallic atoms, sulphur and selenium have a tendency to polymerize, for example, to form atomic chains and rings of the S_n type, where $2 \le n \le 8$ [4.66].

Selenium can exist in several crystalline and amorphous phases, for example, the layered chain structure shown in Fig. 4.12. The atomic structure is different in the bulk material and in thin selenium films: in films, the Se atom chains are shorter, while in the bulk material the chains are long enough to form spheroids. The coordination radius, r_c, and coordination number, N_c, for different selenium phases, vary by $\pm$ 10% [4.67]. The resistivity of selenium also depends strongly on the atomic structure: amorphous selenium has a resistivity of 10^{12} Ohm·cm and crystalline rhombic selenium has a much lower resistivity of $10^5 - 10^6$ Ohm·cm. The latter also depends strongly on impurity concentrations [4.63].

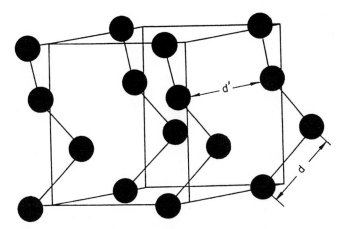

Fig. 4.12. Grey selenium (and tellurium) hexagonal crystal structure. Lattice parameters for Se: $a = 4.3552$ Å; $c = 4.9495$ Å; the minimum interatomic distance in the chains is $d = 2.328$-2.373 Å; the distance between chains $d' = 3.436$ Å. Lattice parameters for Te: $a = 4.45$ Å; $c = 5.91$ Å; $d = 2.86$ Å; $d' = 3.48$ Å

Binary Chalcogenides. The structure of As_2S_3, As_2Se_3 and Sb_2S_3 are typical of binary chalcogenides (Fig. 4.13, 14). The molecular structural unit is tetrahedral AsX_3. Every As atom has five valence electrons ($4s^24p^3$) from which 3 electrons ($4p^3$) can form chemical bonds with X. In the chalcogen atoms X (S $- 3s^23p^4$; Se $- 4s^24p^4$; Te $- 5s^25p^4$) two of the *p*-electrons form chemical bonds and the other two form a non-bonding lone pair [4.69]. As a result, every As atom is covalently bonded to three chalcogen atoms and to each chalcogen, two As atoms (Fig. 3.13).

As_2X_3 has a monoclinic lattice with 20 atoms in the unit cell. The lattice consists of layers made up of rings of 6 AsX_3 units parallel to the surface (010) which are weakly bound through van der Waals forces. The crystal splits easily parallel to the crystallographic axis *a* and *c* , Fig. 4.13. The interaction between the layers is two orders of magnitude weaker than inside the layer. The bulk properties of crystalline As_2S_3 and As_2Se_3 are governed by the two-dimensional layer structure.

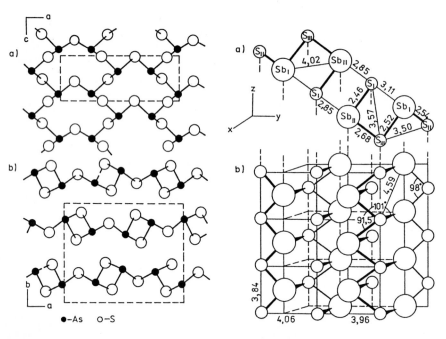

Fig. 4.13. Monoclinic As_2S_3 crystal structure in two projections. Structure parameters: lattice parameters a = 11.46 Å, b = 9.56 Å, c = 4.21 Å; interatomic distance $r_{As-S(1)}$ = 2.15 Å, $r_{As-S(2)}$ = 2.21 Å; valence angles θ_{As} = 102.3°, θ_S = 99.8°

Fig. 4.14a, b. The atomic structure of rhombic Sb_2S_3. Interlayer distances are given in the figure.(**a**) Structure of an $[Sb_4S_6]_n$ layer. (**b**) Layer structure of the bulk material. From [4.68]

Sb$_2$S$_3$ also has a layer structure (Fig. 4.14). Bands of SbS$_3$ units form are parallel to z at an angle of 45° to the x axis and the y axis (Fig. 4.14b). Amorphous Sb$_2$S$_3$ and Sb$_2$Se$_3$ (in contrast to As$_2$S$_3$ and As$_2$Se$_3$) have a strong tendency to crystallize, which can be induced by heating or irradiation. This is because the short range order in amorphous and crystalline Sb$_2$S$_3$ is very similar. The difference between the crystalline and amorphous states, to the first approximation, is that the bands and the chains in the disordered state are shorter. The structure of the amorphous material strongly depends on the technique of preparation.

4.3.2 Electronic Structure

The electronic structure of amorphous semiconductors depends largely on the degree of disorder. Therefore, it is difficult to determine specific values for the band gap, E_g, and the electronic structure of defects for a particular species. In species with a very high degree of disorder, and, therefore, extremely wide electronic states, it is impossible to have local defects similar to point defects in crystals. This has been demonstrated by measurements of the electron conductivity [4.70].

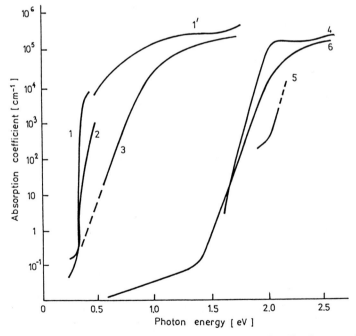

Fig. 4.15. Absorption spectra of Se and Te. Crystalline Te (hexagonal) $E_1 \perp \mathbf{c}$ (*1, 1'*), $E_1 \parallel \mathbf{c}$ (*2*). Amorphous Te (*3*). Crystalline Se: hexagonal (*4*) monoclinic (*5*). Amorphous Se: (*6*)

The extended electronic states can be seen in the broadness of absorption spectra of chalcogenides (Fig. 4.15). Empirical criteria are often used to determine the band gap in that energy region, E_g (or wavelength λ_g), where $\kappa_a(h\nu_g) \approx 10^4$ cm^{-1}, and measurements of the electron conductivity can also be used. The energy band gaps of various crystalline and amorphous chalcogenides are presented in Table 4.6.

Table 4.6. Properties of chalcogenide semiconductors [4.63, 66, 72]

Material	Density g/cm^3	Melting point T_m[a] K	Band gap eV	Refractive index $n(\lambda)$	λ nm
α-As (trig.)	5.73	1090	1.25	–	–
Sb (trig.)	6.69	904	0.12	–	–
α-S (rhomb.)	2.07	392	3.8	$n_1 = 2.0938$	508
				$n_2 = 2.3380$	508
				$n_3 = 1.9876$	508
β-Se (hexag.)	4.79	490	1.6	2.58	1000
Te (hexag.)	6.25	723	0.32	$n_1 = 5.59$	874 ($\perp c$)
				$n_2 = 7.14$	874 (∥ c)
As$_2$S$_3$ (monocl.)	3,48	585	2.6	$n_1 = 3.20$	620
				$n_2 = 2.80$	620
				$n_3 = 2.55$	620
a-As$_2$S$_3$ (amorph.)	3.14–3.20	453 (T_g)	2.4	2.42	820
As$_2$Se$_3$	4.8	633	2.2	$n_1 = 3.24$	632.8
				$n_2 = 2.80$	
a-As$_2$Se$_3$ (amorph.)	4.55–4.62	488 (T_g)	1.8	2.77	1200
As$_2$O$_3$ (arsenolite)	3.89	551 (under pressure)	~2.0	1.755	589.3
As$_2$Te$_3$ (monocl.)	6.23–6.25	658	0.9	–	–
Sb$_2$S$_3$ (rhomb.)	4.64	833	1.9	2.7	1000
Sb$_2$Se$_3$ (rhomb.)	5.84	890	1.0	–	–

[a] The glass transition temperature, T_g, is presented for amorphous materials

The band structure of Se-type semiconductors is formed on the basis of a lattice of sp^3 hybridized Se-atoms. The valence shell of Se is $4s^2 4p^4$, in a p^3 configuration. Thus, two pairs of electrons build σ and lone pair (LP) bonds with neighboring atoms (Fig. 4.16a) and each Se is doubly coordinated. The valence bands are composed of delocalized bonding and lone pair electrons, while the conduction band corresponds to the antibonding σ^* state. The band gap is the energy difference between these two bands. The strong covalent bonding, which is confirmed by the absence of an ESR signal, leads to quite a high energy gap, E_g [4.63, 71].

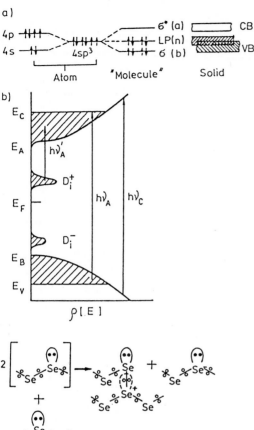

a)

Atom   "Molecule"   Solid

b)

$\rho[E]$

Fig. 4.16a,b. Molecular orbital schemes for the electronic configurations and band structure for Se-like semiconductors. **(a)** Electron orbital energies in a Se atom, Se_2 molecule and Se solid state assuming sp^3 hybridization. **(b)** Density of electronic states for Se-like semiconductors: E_A and E_B are the extended states just above the conduction band (E_C) and below the valence band (E_V); D_i^+, D_i^- are donor and acceptor states; E_F is the Fermi level (the *hatched regions* indicate localized states in which the carrier mobility is low). [4.71]

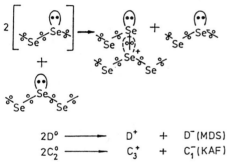

$$2D^0 \longrightarrow D^+ + D^- (MDS)$$
$$2C_2^0 \longrightarrow C_3^+ + C_1^- (KAF)$$

Fig. 4.17. The formation of chalcogenide electron defect states in elemental amorphous Se, corresponding to D^0, D^+ and D^- in the Mott, Davis and Street (MDS) scheme or C_2^0, C_3^+ and C_1^- in the Kastner, Adler and Fritzsche (KAF) model. [4.71]

There are two kinds of bonding defects in amorphous chalcogenides. The first model to describe them was proposed by *Mott, Davis* and *Street* (MDS) [4.73]. This model has also proved successful in explaining a variety of externally induced phenomena in chalcogenides such as photoinduced processes, thermal processes, and luminescence. According to MDS, defects have three charged states D^0, $D^+$and D^- depending on the local coordination. D^0 is a neutral atom, with an unpaired dangling electron, because these atoms are singly coordinated, situated at the end of an Se-chain. D^- are also singly coordinated but they have an extra lone pair. D^+ are triply coordinated sites. D^0 centers are unstable and react exothermically in the charge transfer process

$$2D^0 \rightarrow D^+ + D^-, \tag{4.13}$$

shown in Fig. 4.17. Note that this is not a simple two-center, single-step electron hop, but, rather a two-step process, where the second electron transfer involves creating a D^+ out of a previously undefective site. Figure 4.16b shows a typical energy level diagram for chalcogenide semiconductors, including the energies of the D^- and D^+ defects. Note that the defect energies lie within the energy gap.

A similar model was proposed by *Kastner, Adler* and *Fritzsche* (KAF) [4.74]. This model is somewhat more flexible since it includes more possible combinations of charge states and degrees of coordination. In general, all atoms are designated by C^k_i where k is the charge on the atom and i the coordination number. Thus, a normal Se atom, i.e., neutral and doubly coordinated, is designated C^0_2. Of course, one can easily transform between the MDS and KAF models. Figure 4.17 shows the designations for reaction (4.13) in both schemes. However, in binary or more complex amorphous semiconductors, the KAF model allows one to describe atoms other than chalcogens. A general theory of glassy semiconductors has been proposed by *Klinger* [4.75].

4.3.3 Photoinduced Processes

Photoinduced processes in amorphous chalcogenides were investigated soon after the discovery of this class of disordered materials. Both medium range structure transformations and local structure changes, due to charge transfer, and local bond breaking have been observed. Medium range structural changes take place in metastable amorphous materials (for example, in as-evaporated films) and also in relaxed (thermally or optically) amorphous materials. These changes can be both irreversible and reversible. One of the first investigations of these processes in amorphous As_2S_3 and As_2Se_3 films was performed by *De Neufville, Moos* and *Ovshinsky* [4.76]. In illuminated and thermal annealed arsenic chalcogenide films, they found a strong change in the first and second X-ray diffraction maximum. Both irradiation and heating induce an equilibrium structure in specific difference in the structure of optically and thermaly annealed films [4.63]. Of course, these phase transitions also affect the physico-chemical properties of these species [4.77–79].

Elliot used the EXAFS technique to analyze reversible photostructural changes in annealed a-As_2S_3 [4.63]. He found a stronger change of the second maximum of the radial distribution function during illumination of a-As_2S_3. This demonstrates a change in the medium range order caused by the formation of an interlayer bond. Thus, the concentration of homopolar bonds (As-As, S-S) decreases and the concentration of heteropolar bonds (As-S) increases.

Bishop, Strom and *Taylor* found a broad absorption band in irradiated (at low temperatures) a-As$_2$S$_3$ and a-As$_2$S$_3$ films , with the absorption maximum at approximately $0.5E_g$ [4.81]. These light induced paramagnetic defects with the broad absorption band also lead to luminescence fatigue. They can be thermally annealed at approximately 250 K. No paramagnetic centers are observed to form in arsenic chalcogenides at room temperatures. However, paramagnetic centers are observed in germanium chalcogenides, which have a tetra-co-ordinated Ge disordered network, over the temperature range of 300–400 K.

An example of a medium with a photoinduced phase transition is observed in SmS, for which two stable phases, a metallic (m) and a semiconductor (s), exist [4.60]. The phase transitions s ⇔ m is reversible and correlates with large changes in the coefficient of reflection (Fig. 4.18). However, the use of SmS films as recording media is difficult because of the low reaction rate, even at high temperatures, more than 30 min at T ≈ 600 K, and relatively small light sensitivity (Table 4.5).

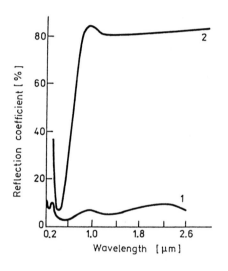

Fig. 4.18. Change in reflectivity due to light induced phase transition from the semiconductor state (s) to the metallic state (m) in SmS films. [4.60]

Another quite interesting characteristic of amorphous chalcogenides is their ability to dissolve in solvents upon irradiation. This process is used in photo- and X-ray lithography as well as for HOE production. *Kenemann* used a 0.01N NaOH solution to photoetch an a-As$_2$S$_3$ sample [4.82]. Etching can be either positive or negative depending on the solution concentration and the rate of solubility of the film. The solubility rate of As$_2$S$_3$ films in a 5% aqueous KOH solution is found to also depend on the exposure time (Fig. 4.19). The contrast can be adjusted for optical, in particular, holographic recording.

Finally, as mentioned above, HOEs (and photo and X-ray lithographs) can be produced by photoetching, yielding elements with high spatial resolution [4.83–85].

Photoinduced processes can be both direct, via electronic excitation which promotes atoms to nonbonding states, in bond breaking and charge transfer, or indirect, when the excited atoms relax by converting the excess electronic energy to vibrational energy (phonons) thereby inducing thermal activated structural changes (short and medium range order). The latter is similar to thermal processes which also affect nuclear motion. The first can occur already at low light intensity, the second usually requires higher light intensities.

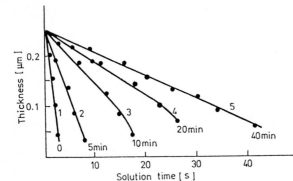

Fig. 4.19. Dissolution of amorphous As_2Se_3 film versus exposure time, illuminated at $\lambda = 488$ nm. [4.85]

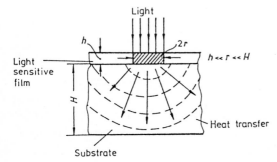

Fig. 4.20. Measurement of light induced thermal heating of thin films of thickness h on a substrate of thickness H: I is the light intensity, r is the radius of the illuminated spot. The condition $h<<r<<H$ must also hold. Direct photostimulated processes depend only on I; photothermal (indirect) processes on $I \cdot r$. [4.86]

These two types of processes in thin films can be easily separated from one another since the rate of the thermal process will depend on the size of the irradiated sample, or rather, the radius of the laser spot, Fig. 4.20, whereas the rate of a direct photoinduced electronic process is size independent. The rate of change of the optical transmission efficiency due to photocrystallization was

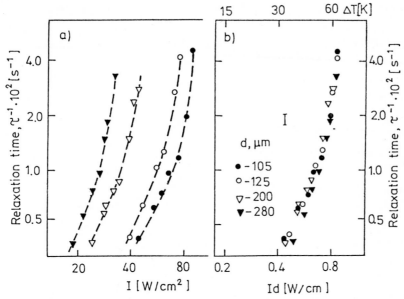

Fig. 4.21a, b. (a) Photocrystallization rate for amorphous Se vs. light intensity for four laser spot diameters. **(b)** Data of part a but plotted vs. $I \cdot 2r$ ($d = 2r$). From [4.86]

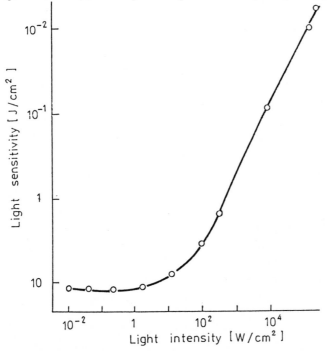

Fig. 4.22. Light sensitivity vs. excitation light intensity in a-As_2Se_3 films

measured for a-Se and a-As$_2$Se$_3$ films [4.86]. It was found that the rate was not only dependent on the laser intensity (20–80 W/cm^2) but also on r (Fig. 4.21), the radius of the spot size, and that the reaction rate $v \sim I \cdot r$. By plotting the Avrami relation $v \sim \exp(-b/nkT)$, where b is the activation energy and n the Avrami factor, we estimated for a-Se films the magnitude of $b/n = 0.52$ eV at $T \leq 345$ K and 0.85 eV at $T > 345$ K. Thermocrystallization data for a-Se from yielded values of 0.78 and 0.85 eV [4.87, 88]. This is conclusive evidence that the structural changes observed in irradiated a-Se (and a-As$_2$Se$_3$) are due to photothermal processes. Presumably, most chalcogenides behave similarly.

Figure 4.22 shows the light sensitivity of an a-As$_2$Se$_3$ film (proportional to the change of the band gap ΔE_g) as a function of the incident light intensity. At low intensities, $I < 10$ W/cm^2, the light sensitivity stays constant whereas above 10 W/cm^2 it is proportional to I and increases by three orders of magnitude. This again confirms that the photoinduced changes in a-As$_2$Se$_3$ at $I > 10$W/cm^2 are due to photothermal, and not direct, processes. Of course, the inducement of higher light sensitivity makes photothermal processes more attractive for optical recording purposes.

4.3.4 Optical Recording

The main applications of chalgonide semiconductor films in optical recording are presented in Table 4.7. The light sensitivity in media used for xerography and thermoplastic recording is much higher than for media treated by direct optical recording techniques, i.e., those which rely on light induced structural changes. Recording by light induced metal atom diffusion (Table 4.5, (3)) is not widely used at present. Photolithography and X-ray lithography (roentgenlithography) in microelectronic and optoelectronic technology are further prospective applications of chalcogenide semiconductor recording media.

Several kinds of photo reactions which affect the optical properties of chalcogenide films can be used for optical recording. These include shifting the band gap, changing the refractive index, or altering the film thickness. It has been found that the magnitudes of these changes are dependent on the stoichiometry. In As$_x$Se$_{1-x}$ the strongest changes are obtained for $x \approx 0.6$ for both as-evaporated and annealed films [4.89, 90]. For as-evaporated films the change of the thickness before and after irradiation $\Delta d/d$ can be 4% while the ratio of the microhardness, $\Delta H/H$, is as much as 700–800% [4.90]. An example of a band gap shift is shown in Fig. 4.23 for as-evaporated and

Table 4.7. Optical recording applications of chalcogenide semiconductors [4.85]

Application	Resolution lines/mm	Sensitivity, S^{-1} J/cm^2	Reversibility
Photoconducting films for xerography	700	10^{-5}	yes
Photoconducting films for thermoplastics	10^3	10^{-5}–10^{-6}	yes
Films on metal substrates	$5 \cdot 10^3$	10^{-1}–10^{-2}	no
Films for optical recording	10^4	1	yes
Films for photolithography	$3 \cdot 10^4$	–	no
Films for X-ray lithography	10^5	–	no

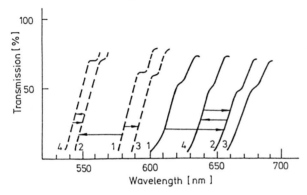

Fig. 4.23. Transmission spectra of amorphous films of As_2S_3 (*solid lines*) and $GeSe_2$ (*dashed lines*); (*1*) as-evaporated; (*2*) after illumination at T_1=300 K; (*3*) after illumination at T_2 = 77 K; (*4*) thermally annealed films. Ref. [4.85]

Table 4.8. Optical anisotropy of chalcogenides

Material	Wavelength nm	Refractive indexes n_1	n_2	n_3	Maximum birefringence
α-S, orthogonal	467	2.012	2.094	2.338	Δn_{31}=0.326
Se, trigonal	1060	3.608	2.790	...	Δn_{12}=0.818
As_2S_3, monoclinic	620	3.200	2.800	2.550	Δn_{13}=0.65
As_2Se_3, monoclinic	632	3.240	2.966	...	Δn_{12}=0.27

annealed films. Irradiation of a film cooled to 77 K, for both As_2S_3 and $GeSe_2$, results in a red shift; irradiation at room temperature results in a red shift for As_2S_3 and a blue shift for GeS_2, also known as "photodarkening" and "photobleaching".

The light induced change in the refractive index in amorphous As_xS_{1-x} (maximum $\Delta n = 0.1$) and As_xSe_{1-x} films (at $x = 0.6$, $\Delta n = 0.78$) are quite high [4.91] and corresponds to a strong change in the polarizability caused by a change in the atomic structure. This light induced change of the polarizability (and corresponding Δn) is similar in magnitude to the natural birefringence of crystalline chalcogenides (which have a high optical anisotropy, Table 4.8). A light induced dichroism and optical activity (i.e., rotation of the plane of polarization) in amorphous chalcogenides was discovered by *Lyubin* [4.127]. The large magnitude of the light induced Δn is useful for phase hologram recording.

4.3.5 Holographic Recording

Chalcogenide films have a high light sensitivity in the visible spectral region and can be easily produced by vacuum deposition or sputtering onto glassy or plastic substrates. Such As_2S_3 films usually have a thickness of 1 - 10 µm and there is no limitation in size.

An effective phase hologram can only be produced in As_2S_3 films by using different wavelength in the recording, e.g. (λ_1), to that used in reconstruction, (λ_2), where (λ_1) $\neq$ (λ_2). This is acceptable for most applications. An exception is coherent optical filtering, where the wavelength for recording and readout must be equal to one an other ($\lambda_1 = \lambda_2$). Therefore, two different holographic recording-reconstruction conditions in a-As_2S_3 films were studied: 1) recording at $\lambda_1 = 514.5$ nm and hologram reconstruction at the same wavelength of

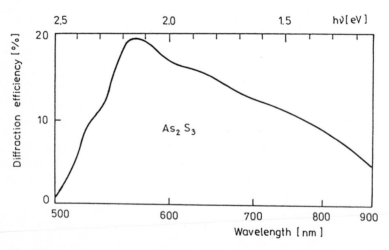

Fig. 4.24. the dependence of diffraction efficiency on the reconstruction wavelength, $\eta(\lambda_2)$, for As_2S_3 films (thickness $d = 6$ µm, recording at $\lambda_1 = 514.5$ nm)

$\lambda_2 = 514.5$ nm ($\lambda_1 = \lambda_2$); 2) recording at $\lambda_1 = 514.5$ nm and reconstruction at $\lambda_2 = 632.8$ nm ($\lambda_1 \neq \lambda_2$). The hologram recording was examined by the kinetics of the efficiency of diffraction $\eta = f(I \cdot t)$ at different intensities (from 0.005 to 2 W/cm^2) and for grating periods, Λ, from 0.3 to 2.5 µm. The light sensitivity was estimated for a 1% diffraction efficiency ($S_{1\%}^{-1}$ [J/cm$^2 \cdot 1\%$], Chap.2).

The spectral dependence of the diffraction efficiency on the reconstruction wavelength is shown in Fig. 4.24 (recording occured at $\lambda_1 = 514.5$ nm). The maximum diffraction efficiency is reached at readout in the spectral region with $h\nu \approx 2.15$ eV (577 nm). By recording at $\lambda_1 = 514.5$ nm, phase hologram properties are dominating read-out in the spectral region from 540 to 780 nm. The recording parameters of As$_2$S$_3$ films are represented in Fig. 4.25, 26 and Table 4.9. The maximum diffraction efficiency, η_{max}, (at the saturation

Table 4.9. Holographic recording in amorphous As$_2$S$_3$ films vs. thickness and light intensity ($\lambda_1 = 514.5$ nm; $\lambda_2 = 632.8$ nm)

Thickness, d µm	Grating period, Λ µm	Light intensity, I W/cm^2	Exposure, It J/cm^2	Exposure time s	Maximal diffraction efficiency $\eta_{max}\%$	Light sensitivity S^{-1} J/cm$^2\%$
Recording and readout with $\lambda_1 \neq \lambda_2 = 632.8$ nm						
11.2	1	0.2	14	72	76	0.18
		2.0	30	15	67	0.45
5.2	0.7	0.2	19	96	73	0.26
	1.6	15	24	56	0.27	
2.6	0.7	0.2	24	120	28	0.86
		1.6	24	15	25	0.96
1.3	1	0.2	19	96	9.0	2.11
		1.6	18	11	8.5	2.11
Recording and readout with $\lambda_1 = \lambda_2 = 514.5$ nm						
11.2	1	0.2	9.6	48	5	1.92
	1.6	16	10	4	4.00	
5.2	0.7	0.2	9.6	48	10	0.96
	1.2	15	12	6	2.50	
2.6	0.7	0.2	19	96	11	1.72
1.3	0.7	0.2	19	96	8	2.37

exposure) depends on the thickness of the sample and the light intensity. For thick samples ($d = 11$ μm), the saturation value, η_{max}, is larger at lower intensities and the saturation takes place at smaller exposures. The reciprocity law, $I \cdot t$ (Chap. 2) does not hold for $\lambda_1 = \lambda_2$ and $\lambda_1 \neq \lambda_2$ at $I \geq 0.8$ W/cm². The exposure for a-As$_2$S$_3$ films of thickness $d = 5$ μm at which saturation occurs for different light intensities are similar (Fig. 4.25), however, the saturation values, η_{max} are remarkably different. This can be explained by the dependence of the photoreaction kinetics on the intensity of the light. The recording process leads to a shift in the band gap (photodarkening), which also influennces the change in light absorption during illumination. Such dependence of the recording parameters on the light intensity is typical for many chalcogenides.

Readout at the same wavelength as the recording wavelength may cause destructive reconstruction of the hologram. For this reason, the influence of the light intensitiy on hologram reconstruction was analyzed. A critical intensity, I_{cr}, could be found below which the destruction of the recorded hologram during readout is negligible. For a-As$_2$S$_3$ films at $\lambda_1 = \lambda_2 = 514.5$ nm, $I_{cr} = 0.01$ W/cm².

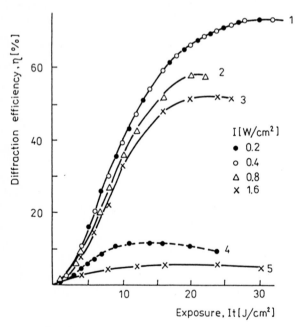

Fig. 4.25. Holographic recording kinetics, $\eta(I \cdot t)$, in a-As$_2$S$_3$ films (thickness d=5.2 μm; grating period Λ=0.7 μm) vs. light intensity for $\lambda_1 \neq \lambda_2$ (1, 2, 3) and for $\lambda_1 = \lambda_2$ (4, 5); λ_1=514.5 nm; λ_2=632.8 nm

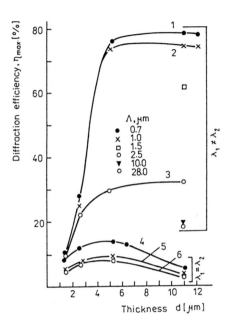

Fig. 4.26. The influence of the film thickness d and grating period Λ on the maximum diffraction efficiency of As$_2$S$_3$ films (for $\lambda_1 = 5.14.5$ nm $\neq$ $\lambda_2 = 632.8$ nm and $\lambda_1 = \lambda_2 = 514.5$ nm): curves *1, 2, 3* for $\lambda_1 \neq \lambda_2$; *4, 5, 6* for $\lambda_1 = \lambda_2$. The total light intensity ($I = I_S + I_R$) was 0.2 W/cm^2

If reconstruction of a recorded hologram at λ_1 514.5 nm occurs at a different wavelength $\lambda_2 = 632.8$ nm, the Bragg angle correction for the reference beam must be used (Chap. 2). Such particularities in diffraction were observed in amorphous As$_2$S$_3$ films, with the thickness $d \geq 5$ µm, by recording and readout at different wavelengths, i.e. $\lambda_1 \neq \lambda_2$ ($\eta_{max} = 78\%$; Figs. 4.25, 26). Such results, including reflection losses (approximately 20%), are close to the theoretical limit $\eta_{max} = 100\%$, as one can come.

As mentioned in Chap. 2, the ratio, Λ/d determines the type of the hologram, whether it is a thick (volume, 3-D) or a thin (2-D) hologram. Volume holograms can only be produced in As$_2$S$_3$ films with a thickness of $d \geq 10$ µm for holographic gratings with period $\Lambda \leq 2$ µm. Under these conditions, the recorded holograms under the Klein criterion (Chap. 2, (2.9)) have Q > 10 and are of volume (3-D) type. The results in Fig. 4.26 demonstrate that high diffraction efficiencies can also be reached in As$_2$S$_3$ films with a smaller thickness. However, if the film thickness (for the same grating periods $\Lambda \leq 2$ µm) is smaller than 10 µm, the value Q < 10 and the recorded hologram is not a 3-D hologram. According to the Klein criterion (Chap. 2, (2.9)), holograms in As$_2$S$_3$ films, with thickness 2 µm $\leq d \leq 10$ µm, are intermediate holograms with a high diffraction efficiency but a much smaller angular resolution than for 3-D holograms (Sect. 2.4.4, Fig. 2.8).

Coated As_2S_3 films are more common for practical applications. The coating prevents the As_2S_3 film from mechanical destruction and interaction with the atmosphere. The coating can also play the role of an anti-reflection layer, when it has a definite thickness (the interference conditions for anti-reflection layers can be easily estimated [3.4]). Good results were obtained with SiO_x and MgF_2 coatings for As_2S_3 films, having good mechanical and anti-reflection properties [4.80].

Hologram Self-Enhancement. Hologram self-enhancement (HSE) was discovered by *Staebler* and *Amodei* by phase recording in $LiNbO_3$ crystals [4.49,50]. This was observed by *Shvarts*, *Ozols* and *Reinfelde* [4.92] in As_2S_3. HSE occurs in 3-D holograms by recording with two beams (I_S, I_R) up to a given diffraction efficiency, η_0 (Fig. 4.27a). By a subsequent illumination or dark reactions, an increase in the diffraction efficiency was observed. The HSE is described by the self-enhancement coefficient defined as:

$$\xi = \frac{\eta(t)}{\eta_0}. \tag{4.14}$$

Three different HSE phenomena were observed in As_2S_3 films (Fig. 4.27): (1) coherent; (2) non-coherent and (3) dark.

Coherent HSE occurs in 3-D holograms by illuminating with a reference beam I_R (usually at the same angle of incidence, i.e., at Bragg conditions). Thus, the reference beam is scattered (diffracted) by the holographic grating, the hologram generates a scattered subject beam, I_S, which causes interference and further recording with two beams takes place. The coherent HSE depends on the initial diffraction efficiency, the dynamic range of the holographic recording (i.e., the maximum values of the light induced optical constant changes Δn, Δk) and the wavelength of recording and readout. The HSE is much stronger for phase holograms than for amplitude holograms. We observed a strong HSE for small initial diffraction efficiencies $\eta_0 = 10^{-5}\%$ in As_2S_3 films. The maximum self-enhancement coefficient, ξ, for phase holograms (recording at $\lambda_1 = 514.5$ nm; readout at $\lambda_2 = 632.8$ nm) is about 1000 times larger than in $LiNbO_3$. Such effective self-enhancement in As_2S_3 films leads to the strong change in the polarisability under illumination and the corresponding large dynamic range for phase recording. The change of the refractive index, Δn, in As_2S_3 films is approximately $\Delta n \approx 0.1$, in $LiNbO_3$ crystals $\Delta n_{max} \approx 3 \cdot 10^{-3}$ (Sect. 4.2, formula (4.9)). This large difference in the change in the light induced refractive index explains the difference in the magnitude of the self-enhancement [4.92,93].

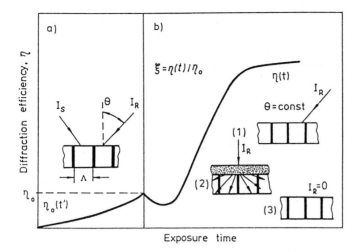

Fig. 4.27a, b. Hologram self-enhancemt (HSE): **(a)** Holographic recording with two beams I_S, I_R up to a diffraction efficiency, η_0; **(b)** (*1*) coherent HSE under Bragg conditions; (*2*) non-coherent HSE by excitation with diffuse scattered light; (*3*) HSE by dark reactions

By recording and readout at the same wavelength ($\lambda_1 = \lambda_2$) the HSE is smaler [4.93]. For $\lambda_1 = \lambda_2 = 514.5$ nm, the maximum HSE factor is $\xi \leq 10$. The HSE decreases at higher exposures, when the initial η_0 is close to η_{max}. This means that the recording process is saturated and no further recording is possible. The HSE also depends on the holographic grating period. The maximum value of ξ corresponds to holographic gratings with a period of 1-2 μm. Thus, the holographic self-enhancement is typical for 3-D holograms and is absent in 2-D holograms (ξ approaches to zero in films with a small thickness or for large holographic grating periods).

Non-coherent HSE takes place by illumination with scattered light (Fig. 4.27b (2)). In this case, the 3-D hologram acts as a spatial filter which selects the light beams at the Bragg angle. Then, HSE occurs, as discussed above. However, for non-coherent HSE, the self-enhancement factor, ξ, is much smaller (for one or two orders) than for coherent HSE. Coherent and non-coherent HSE is a general phenomenon which occurs in any real time recording medium. This effect must be taken into account in real time optical recordings [4.19, 93].

Dark hologram enhancement is typical for As_2S_3 and some other chalcogenide films. This phenomenon is caused by the recording mechanism in As_2S_3 films. Illumination of As_2S_3 films induces a change in the medium range order and interlayer bonding [4.90, 91]. Thus, the illuminated film is denser and thinner, and the holographic recording induces a mechanical stress with a periodical distribution (with period Λ of the recorded grating). In darkness,

Table 4.10. Example of photoinduced processes in organic materials

Reaction type	Example
	Anthracene[a]

(250 nm; 350 nm)   (220 nm)

Salicylidenaniline

(340 nm)   (400-500 nm)

Stilbene

cis (280 nm)   trans (480 nm)

2, 4, 5-phenyl-3-R-cyclo-pentadienoxide

(400 nm)   (550 nm)

Reaction type	Example
Oxidation-reduction	
Proton tautomerization	
cis-trans isomerization	
Valence isomerization: redistribution of π- and σ-bonds	

Table 4.10. (continued)

Reaction type	Example

6-Phenoxy-5, 12-naphthacenquinone

Rearrangment

Ph–O

(395 nm)   (450-490 nm)

Tetrachloro-ketonaphtalene

Homolytic bond cleavage

(400 nm)   (525 nm)

Thymine

Polymerization

(270 nm)   (240 nm)

[a] Molecular names refer to reactant species; the absorption maximum is given in parantheses

the stress field can relax, allowing self-enhancement. The dark relaxation time is several days and the HSE is of the order of 5-10. This self-enhancement effect is not a general effect and occurs only in some special cases.

HSE can be used for optical recording at low light intensities with subsequent enhancement. However, this enhancement effect depends on the grating period, Λ. Therefore, the HSE of a real image (i.e., a superposition of different interference structures) will distort the image. For the depression of the HSE in real-time optical recording, the diffraction efficiency of the recorded hologram must be close to the maximum (saturation) value.

4.4 Organic and Polymer Materials

It is easier to vary the optical properties of organic compounds than of inorganic materials. This can be done by stimulating electron and proton transport, tautomerization, isomerization, or polymerization [4.94, 96]. The optical properties of a given molecule (spectral bandwidth, light sensitivity) can easily be changed by placing it in different polymer matrices (e.g., polymethylmethacrylate (PMMA), polyvinylchloride) which is common in many applications. Recently, the interest in organic recording materials is directed to holographic memory systems (2-D and 3-D) and holographic interconnections for neural networks [4.97–99]. Organic films are more common for optical and holographic recording than crystalline materials (no size limitation, simple preparation technology and greater cost-effectiveness). Additional advantages of organic recording materials are their high optical nonlinearity, which opens new fields of application in laser techniques and nonlinear optics [4.100, 101]. Different dye-polymer films and various recording methods are examined for read-write-erase holographic memory systems (one and two photon excitation, time and frequency domain holography) [4.97, 99, 100]. However, all these developments are only at an initial stage of application. Table 4.10 shows examples of the most commonly observed types of photoinduced processes in organic compounds. A brief description of some organic recording materials follows.

4.4.1 Camphorquinone

Among the first organic species to be tested as a potential holographic media were the quinones, in particular, camphorquinone (Table 4.11) in PMMA [4.103]. Illumination with 488 nm Ar^+-laser light creates a free radical (probably a light induced proton transition from one of the CH_3 groups to an oxygen) which is quite efficiently stabilized by the matrix. This is, of course, useful for long term data storage. However, the light sensitivity is low, 1 J/cm^2, unless additional heating of the matrix, up to ~ 340 K, which leads to an increase in the diffraction efficiency, up to 1000 times (Fig. 4.28).

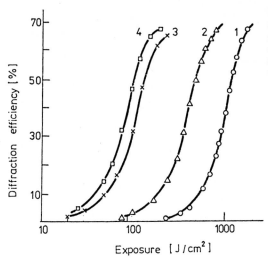

Fig. 4.28. Dependence of the diffraction efficiency of camphorquinone in polyacrylate on the exposure intensity $I(t)$. Camphorquinone con-centrations (% mass): (*1*) 1%, (*2*) 3%, (*3*) 5%, (*4*) 10%. [4.103]

Table 4.11. Parameters of organic solid solutions in benzoyl peroxide $(C_6H_5CO)_2 \cdot O_2$ used as holographic recording materials [4.103]

Material	Concentration mass % nm	Recording wavelength %	Diffraction efficiency temperature	Exposure J/cm^2	Storage time at room
2,5-dimethyl-3,4 -hexadione	5	488	75	60	3 weeks
Camphorquinone $(C_{10}H_{14}O_2)$	5	514.5	70	1000	Stable after rapid decrease
Thymoquinone $(C_{28}H_{30}O_4)$	0.25	514.5	5	400	stable
β-Bromo-β-nitro-styrene $(PhCH=C(Br)NO_2)$	5	488	5	300	3 days

4.4.2 Bacteriorhodopsin

A number of research groups have demonstrated the reversible optical storage [4.104 – 107] and nonlinear properties [4.100, 102] of bacteriorhodopsin. Bacteriorhodopsin (bR) is the light transducing protein in the purple membrane of the Halobacterium halobium [4.98]. This bacterium thrives in salt marshes, where the concentration of sodium chloride is approximately six times greater than in sea water. Bacteriorhodopsin can be easily produced by these bacteria.

For practical applications, bacteriorhodopsin can be incorporated into various water soluble polymers (polyvinyl, bovine gelatin, polyethylene glycol, methyl cellulose etc.). By addition of chemicals (guanide hydrochloride [4.108], arginine [4.109], polyvalent metal salts [4.110]) the recording properties of bR can be varied. Rhodopsin films produced in this manner are attractive for applications.

Optical recording in bR films is reversible. As mentioned above, bR-polymer films are used for optical recording . The film thickness varied from 30 to 500 μm and the optical density depends on the concentration of bR in the polymer matrix.

Bacteriorhodopsin is a reversible recording medium without longterm information storage. The recording mechanism is a photoconversion of bR from the initial IbR to metastable MbR states:

$$IbR \Leftrightarrow MbR .$$

(4.15)

The absorption maximum of IbR occurs at $\lambda_1 = 568$ nm and the photoconversion induces the MbR state with the absorption maximum at $\lambda_2 = 413$ nm. Under standard biological conditions, the MbR state reverts to the ground state (i.e., initial state IbR) in a time of approximately 10 ms. The lifetime of the metastable state MbR can be strongly increased to 15–30 minutes in polymer matricies with the addition of chemicals. This is sufficient for optical processing but not acceptable for long-time storage, e.g., archivale memory. The direct photoconversion can be induced in the spectral region 480–620 nm and the reverse reaction MbR $\rightarrow$ IbR (in addition to thermal reversion) in the region 350–450 nm. The reaction (4.15) has a relaxation time of $\tau \approx 50$ μs. Amplitude and phase readout is possible in both states (IbR, MbR). At high bacteriorhodopsin concentrations (30 weight %), the optical density (D) for a 150 μm thick bR-polymer film is $D \approx 1.3$ (in both states IbR, MbR). Therefore, amplitude recording is only possible in the strong absorption region . However, high efficiency phase recording for IbR $\Leftrightarrow$ MbR can be accomplished at wavelengths of $\lambda \leq 480$ nm and $\lambda \geq 620$ nm ($\eta_{max} \approx 11\%$)[4] [4.98]. The reversible photoconversion in bR (4.15) can be used in holographic memory systems and as an optically addressed spatial light modulator. Those recording parameters of bR which have been attained are shown in Table 4.12.

---

[4] The maximum value of η for bR is much lower than for As_2S_3 film ($\eta_{max} \approx 80\%$). Chalcogenide films (As_2S_3, As_2Se_3) are unique materials with extremely high light induced refractive index changes (Sect. 4.3.5)

Table 4.12. Recording properties of thin polymeric bacteriorhodopsin films [4.98]

Reaction	IbR → MbR	Mbr → IbR
Recording wavelength range nm	480–620	350–450
Readout wavelength range nm	350–820	350–820
Quantum yield	≥ 0.64	0.64–0.95
Minimal recording time in μs	≤ 50	≤ 50
Light sensitivity mJ/cm^2	≤ 100	≤ 100
Thermal relaxation at $T = 300$ K in min	~15	~15
Diffraction efficiency in %	10%	10%
Photocyclicity number of cycles	> 10^6	> 10^6

4.4.3 Reoxan

This material can be used for three dimensional phase recording. The thickness is typically between 0.2 and 3 mm; for thinner films, a glass substrate is required. The photoactive component is usually anthracene, in concentrations of 0.04–0.3 mol/l which, because of it´s transparency to UV and visible light, requires a photosensitizer, i.e., an organic dye with a concentration of $5 \cdot 10^{-6}$–10^{-4} mol/l, depending on the absorption cross section [4.111]. This two-component PMMA matrix must be chemically synthesised with molecular O_2, which is allowed to diffuse slowly through the medium. A PMMA matrix with thickness 1 mm and subjected to a pressure of between 50 and 140 atm, requires two to three days to be synthesized (Fig. 4.29a).

A cyanine dye – quinolinum chloride (absorption maximum at 580 nm) is used for optical recording with green light (500–550 nm). The dye molecule absorbs the photon and is excited to a triplet electronic state (Fig. 4.29b) and the absorbed energy is transferred (by non-radiative energy transfer) to the oxygen molecules promoting them to a chemically active excited state:

$$D^* \to O_2\left(^3\Sigma\right) \to O_2^*\left(^1\Delta\right), \tag{4.16}$$

where D^* and O_2^* are the excited of the dye and the oxygen molecules and $^3\Sigma$ and $^1\Delta$ are the triplet ground and first excited singlet states of the oxygen molecule. The excited oxygen O_2^* can react with anthracene:

$$A + O_2^* \rightarrow AO_2. \tag{4.17}$$

Reaction (4.17) is illustrated in Table 4.10. After recording, the material is fixed by allowing the excess oxygen, not chemically bound to anthracene, to rediffuse out of the medium. This takes about two weeks (at room temperature) for a 1 mm thick material.

The oxidized species has a different index of refraction. For example, exposure to 1 J/cm^2 in the spectral region of 440–514.5 nm (He-Cd or Ar$^+$-lasers) yields $\Delta n \approx 0.02$, which in the transparent PMMA matrix gives a diffraction efficiency for holographic recording close to 100%.

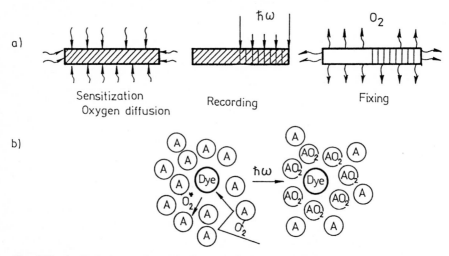

Fig. 4.29a, b. Optical recording using Reoxan. **(a)**Three steps of the recording process. **(b)** On the left we show the impinging photon exciting a dye molecule in a Reoxan matrix. An O_2 molecule collects the energy from the dye and O_2^* is incorporated into the anthracence molecules A. [4.111]

Two parameters of the dye must be taken into account for optimal recording in Reoxan: (1) the absorption efficiency must be sufficiently high in the spectral region of the recording wavelength; (2) the energy of the excited state, in the dye molecule, D^*, must be higher than (or equal to) the activation energy in the oxygen molecule, O_2, (the transition energy $^3\Sigma \rightarrow ^1\Delta$ is approximately 1 eV). The energy of the excited state of the molecule in numerous dyes, e.g. cyanine and Rhodamine 6G, suffices to drive the transition from the triplet to the single configuration in the O_2 molecule.

Thus, by using different dyes, it is possible to make an optical recording in Reoxan over the entire visible, and near the infrared, spectral region. The high diffraction efficiency (up to 100%) in the transparent PMMA matrix is quite a positive property for high quality holographic recording, however, the long processing time and low light sensitivity do limit its attractiveness.

4.4.4 Thermoplastic Recording Media

A relief hologram in thermoplastic recording is created by modulating the recording medium thickness, $d(x, y)$. Recording on these materials is a three-step process, shown in Fig. 4.30. In the first step, a conductive substrate is uniformly charged by a corona discharge in air (Fig. 4.30a). In the second step the medium is exposed to light which changes the charge distribution

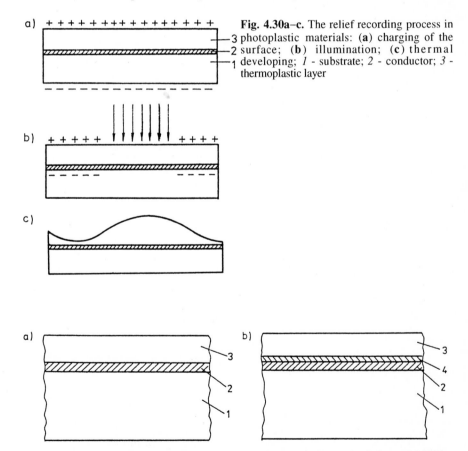

Fig. 4.30a–c. The relief recording process in photoplastic materials: (a) charging of the surface; (b) illumination; (c) thermal developing; 1 - substrate; 2 - conductor; 3 - thermoplastic layer

Fig. 4.31a, b. Thermoplastic recording systems. 1, substrate, 3, thermoplastic layer. (a) With a conductor layer 2. (b) With a separate conductor 2 and photoconductor 4. [4.112]

in the irradiated regions (Fig. 4.30b). The third step involves thermal developing (IR radiation, contact or radio frequency heating) of the latent electrostatic image producing a surface relief (Fig. 4.30c). Usually, the recording time (including thermal fixation) is of the order of 1 ms to 10 s. The three recording steps can be carried out separately, or simultaneously.

Several types of photothermoplastic recording materials are available: with a single conductive layer or with separate conducting and photoconducting layers (Fig. 4.31) [4.112]. More complicated layering can result in higher light sensitivity. The photoconductive layer acts as the charge carrier generator in the charge redistribution step, under illumination. The purpose of the photoconductor is similar to that of the synthesising dye in Reoxan systems. Table 4.13 lists typical parameters of photothermoplasts. It has been found that a separate photoconducting layer (Fig. 4.31b) injects more charge carriers than a photoconducting thermoplastic material. This yields better image quality and a higher degree of amplification of the initial latent image (i.e., before thermal developing). The total image amplification is of the order of 10^4–10^7. The sensitivity of the medium also depends on its conductivity which determines the electric field gradient and the ensuing photoconductivity.

Table 4.13. Recording parameters of photothermoplastic materials

Material	Wavelength nm	Light sensitivity, S^{-1} for $\eta=1\%$ J/cm^2	Maximum diffraction efficiency η %	Exposure time s	Enhancement factor
Polyvinyl-N-carbazole-Se	530	$2 \cdot 10^{-7}$	15	10^{-8}	10^7
Polyvinyl-N-carbazole-CdSe	694	10^{-8}	20	10^{-8}	10^7
Polyvinyl-N-carbazole-trinitrofluorenone	633	$2 \cdot 10^{-5}$	20	1	10^5

Thermoplastic materials up to several microns thick can be used as two-dimensional recording materials. The maximum diffraction efficiency for transmission phase recording is 33.9% (without reflection losses). Higher diffraction efficiencies can be reached by additional metallization after production of relief (that is, creating reflection holograms).

The erasure of the profile image is accomplished by heating up to 100–150° C, i.e., to the glass transition temperature, T_g. Usually, up to $5 \cdot 10^4$ annealing cycles are possible [4.112].

Typically, polyvinyl-N-carbazoles with thickness of up to 1 μm are used in thermoplasts. For conducting layers, transparent In_2O_3 or SnO_2 with thickness of $d < 0.1$ μm are used. The photoconducting injection layers are organic dyes or inorganic semiconductor films (Se, CdS, Te, Ge), 0.1 – 1 μm thick.

Commercial photothermoplastic materials have a high spatial resolution of 3000 lines/mm. In the best materials, the signal-to-noise ratio is 40 dB (for smaller relief amplitudes the SNR decreases down to 30 – 35 dB). The light sensitivity, taking into account the high amplification factor (10^7), is of the order of $10^{-5} – 10^{-7}$ J/cm^2 (for $\eta = 1\%$) which is comparable with, or higher than, silver halide materials.

Thermoplastic recording materials have a wide region of application in holography and laser interferometry. Their only limitation is in real time information storage, because of the required thermal processing.

4.5 Persistent Spectral Hole Burning

The absorption spectrum of many color or impurity centers in condensed matrices (crystals, glasses, solid solutions) is actually an inhomogeneously broadened superposition of many narrow homogeneous lines of width δ_0. The frequencies of these lines are different from one another, even for the same chemical species, because of the differing local environments or the orientation of the local centers (dipolar molecules, atomic clusters etc., Fig. 4.32a). By using a very narrow band laser, it is possible to slowly tune over the absorption peak and select out the individual lines. When the laser frequency exactly matches a particular absorption frequency, a strong dip, a spectral hole, is burned into the peak (Fig. 4.32b): the laser excitation depletes the number of particles in that absorbing state, or, reorients the local center (dipole). If this laser induced state is very long lived, that is, if the transition corresponds to a physical transformation of the irradiated area, then it can be used for optical storage (digital or holographic). This is called persistent spectral hole burning (PSHB).

4.5.1 Frequency Domain Optical Storage

The spectral hole can be burned at a number of different frequencies (Fig. 4.32c). Such frequency domain optical storage, in which the presence of a spectral hole is coded "1" and its absence "0", allows very high data storage densities. The storage density is limited by the ratio of the homogeneous width,

δ_0, of the individual lines to the inhomogeneous width, δ_a, (Fig. 4.32a). Analog storage is also possible and the high storage density can be used for optical imaging or holographic recording [4.113]. Usually, PSHB is possible only at low temperatures (1–4 K).

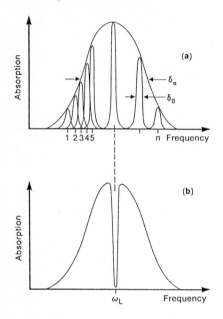

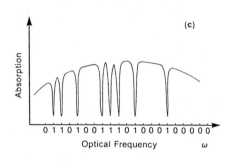

Fig. **4.32a–c.** Frequency domain data storage using persistent spectral hole burning (PSHB): (**a**) inhomogeneously broadened absorption line; (**b**) absorption after spectral hole is burned by illumination with a narrowband laser at ω_L; (**c**) multiple hole data storage in the frequency domain. [4.113]

Rebane et al. studied the absorption spectrum of H_2 - phtalocyanidin in n octan matrix at 4.2 K [4.114]. This substance was found to undergo persistent hole burning by photoinduced 90° rotation of the H_2 molecule in the matrix [4.115]. The linewidth of the burned hole was $\delta_0 \approx 10^{-3}$ cm^{-1} ($3 \cdot 10^7$ Hz) is much smaller than the bandwidth of the absorption spectrum δ_a (Fig. 4.32a). In optimal matrices, the ratio $\delta_a/\delta_0 \approx 10^5$–$10^6$. The maximum storage density for a 2-D medium is $N_{2D}^{max} \approx 10^8$ bit/cm^2 (for $\lambda = 500$ nm, Chap. 2)). On any diffraction limited spot (with the area $\Delta S \approx \lambda^2$), we can store, by PSHB additional $N_{PSHB} = \delta_a/\delta_0$ bits. Thus, the total (N_{tot}) storage density for PSHB is:

$$N_{tot} = N_{2D}^{max} \cdot \frac{\delta_a}{\delta_0}. \qquad (4.18)$$

Since, in typical systems, 1 bit of data requires approximately 100 holes, the final effective storage capacity, according to (4.17), is 10^{12} bit/cm^2.

The high storage density, however, limits the recording time, τ, and the data transfer rate. From the uncertainty relation, the linewidth of the burned hole, δ_0, and the recording time, τ, are determined by: $\delta_0(v) \cdot \tau \geq 1$. Thus, for the linewidth: $\delta_0(v) \approx 3.10^7$ Hz (10^{-3} cm^{-1}), we arrive at minimum recording time for one bit as $\tau \geq 3.10^{-8}$ s. This means that for the maximum storage density of 10^{12} bits/cm^2, 8 hours are required for single-beam recording (or readout). The long recording and reading times are not the only unfortunate disadvantages of this, otherwise, quite interesting method. The primary drawback is the requirement of recording at liquid helium temperatures. Note that these are needed, not only to be able to observe the lines, but, also to prevent annealing, or other reactions, in the vicinity of the holes. Liquid helium refrigerators are often more expensive than any optical disc system.

Concluding, we would like to mention the work of *Winnacker* et al. who observed PSHB in a two-photon hole burning process in a BaClF-Sm^{2+} crystal [4.116]. The holes are stable at room temperature; cooling to 1-2 K is required only to resolve the spectral hole $\delta_0 = 8.3 \cdot 10^{-4}$ cm^{-1}.

S. Arnold et al. [4.117] tested the PSHB method in polystyrene microparticles. The random particle distribution (with sizes of the order of 1-2 µm) allows PSHB (recording and readout) at room temperature. *Schellenberger, Lenth* and *Bjorklund* discussed the technological aspect of the data storage using persistant spectral hole burning [4.113]

4.5.2 Time-Domain Holography

Spectral hole-burning spectroscopy has also stimulated the development of dynamic time-domain holography. Figure 4.33 illustrates the principles of this holographic recording by two light pulses, $S(\mathbf{r}, t)$ (with pulse length τ_s) and $R(\mathbf{r}, t_R)$ (with pulse length τ_R) which reach the recording medium at different moments, $t = 0$ and $t = t_R$. The first pulse is the object (or subject) beam while the second, $R(\mathbf{r}, t_R)$, is the reference beam. If the pulse lengths τ_S, τ_R and the delay time, t_R, are shorter than the phase relaxation time, $\tau_2 \sim 1/\delta_0(v)$ (usually τ_2 is on the order 10^{-7} s), the interference of these two pulses records a hologram which is called a time-domain hologram. Such a hologram is recorded at different moments of time: the recording medium stores the phases and amplitudes of the first pulse.[5]

The idea of time-domain holography was first discussed by *Mossberg* [4.118]. A general description of this phenomenon is given in [4.115]. The first experiments on time-domain holography were performed by *Rebane, Kaarli* and *Saari* with H$_2$-tetra-tri-butyl-porphyrazine in a polystrene matrix by excitation with Ar-ion pumped tunable dye laser; pulse length 2-3 ps at 82 MHz

---

[5] The time limitation determinated by the uncertainty relation is absent by holographic recording, i.e., in time-domain holography [4.122] (Sect. 4.5.1)

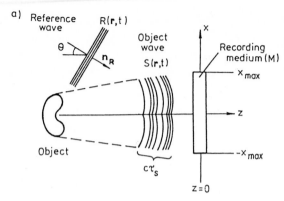

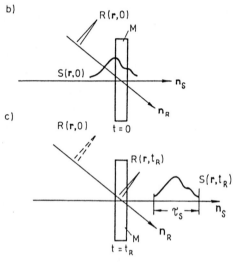

Fig. 4.33a–c. Time-domain holographic recording. (**a**) The object beam, $S(\mathbf{r}, t)$, and the reference beam, $R(\mathbf{r}, t)$, reach the recording medium at different times. (**b**) The recording medium stores the exciation from $S(\mathbf{r}, 0)$ and interacts with the excitation of the reference beam $R(\mathbf{r}, t_R)$. (**c**) Readout of the time-domain hologram with the pulse $R(\mathbf{r}, t)$; the output signal is the subject beam $S(\mathbf{r}, t_R)$

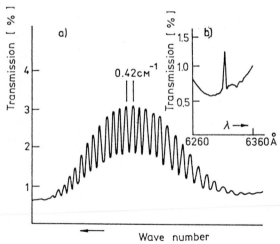

Fig. 4.34 (a) A spectral time-domain hologram in the transmission spectrum of H_2-tetra-tri-butyl-porphyrazine in a polystyrene matrix, recorded at $T = 1.8$ K with 2–3 ps dye laser pulses (82 MHz repetition rate). (**b**) The *insert* shows the spectral hole at low resolution (~ 0.1 nm). [4.120]

repetition rate [4.119]. A spectral time-domain hologram is shown in Fig. 4.34 a.

Time-domain holography is a new method for spatial-temporal optical processing. This method gives new possibilities for high resolution spatial filtering. This method was used in [4.120, 121] for fast time process analysis and summation and substraction of optical signals. Some new aspects of crosstalk minimization in time-domain holography by femtosecond excitation are discussed in [4.122]. By a similar technique, *Kachru* et al. recorded 100 bits on a single micron sized spot in $YAlO_3$-Pr^{3+} crystals [4.102]. Persistant hole burning and time-domain holography are novel optical recording methods for future memory systems [4.124, 125, 126].

5. Digital Recording
   on Optical Memory Discs

The basic principles of reversible and nonreversible photothermal recording as applied to commercially available optical memory discs are described in this chapter. Two types of recording materials exist, phase-change and magneto-optic. In the first, the heat resulting from exposure to light induces a phase transition (atomic structure changes) which changes the transmission properties of the medium. In the second, light induces a change of the magnetic order, which alters the local magnetization of the material; which can be detected via the Faraday or Kerr effects discussed in Chap. 3. The magnitude of the primary optical signal change in magneto-optic (MO) materials is smaller than in phase-change (PC) media. In addition, the former require a more complicated optical reading device for light polarization analysis, and are, thus, more costly. Both types of disc media are important and can be used in applications where eraseability is desirable. The technology of optical memory discs is analyzed in detail by *Marchant* [5.1]. Thus, only a brief description of the optical storage devices is given here. In Sect. 5.3, new luminescent recording systems are discussed.

5.1 Commonly Available Optical Memory Discs

Optical discs have been commercially produced since 1982. Their main applications have been, so far:

- audio and video recordings
- personal computer memory
- archival memory.

Discs come in standard sizes with diameters of 130, 200, 305, and 360 mm, of which the 130 mm is the most common. The storage density ($\sim 10^8$ bit/cm^2) is only several times larger than that of the best magnetic discs. However, the cost per bit is one to two orders of magnitude lower than that for magnetic recording [5.2]. The storage density of a 130 mm disc, $3 \cdot 10^9$ bits/side, is enough to store 50 000 pages of printed text, or a 30-minute video program (54 000 frames with sound), or a 70-minute audio program.

The parameters of MO and PC recording materials are similar (Table 5.1). Both types of media can be used for reversible recordings and WORM (write-once-read-many) applications [5.2–5]. The difference between ROM

Table 5.1 Specifications for erasable optical discs using continuous-composite format [5.2]

Disc Format

Radius of that surface available for use	30–60 cm
Available tracks per side	18 751
Sectors per track	17 (1024 byte/sector)
	31 (512 byte/sector)
Track configuration	Spiral
Track pitch	1.6 μm

Capacity (two-sided)

per unformatted side	434 Mbytes
per formatted side (1024 byte/sector)	325 Mbytes (1024 byte/sector)
	298 Mbytes (512 byte/sector)

Mechanical Specifications

Disc diameter (exterior)	130 mm
Cartridge dimensions	135 mm / 153 mm / 11 mm

Reliability

Read cycles	$>10^7$
Erase/write cycles	$>10^6$
Archival life (accelerated testing)	>10 years
Shelf life (accelerated testing)	>10 years

Environment

Operating temperature	10/50 °C
Operating humidity	10–80%
Storage temperature	–10/55 °C
Storage humidity	10–90%

(read-only-memory) and WORM is that the latter are capable of real time recording [5.1, 2]. The photophysical and photochemical processes by which recordings are made have been discussed in Chap. 4. Nonreversible recordings generally involve open hole or bubble creation, whereas reversible recordings rely on phase transitions or chemical reactions which change the optical constants, κ and n, or the magnetization (Fig. 5.1).

5.1.1 Multilayered Recording Media

Figure 5.1 and Fig. 5.2 show how the discs are composed of several layers: a thick substrate (polymethyl-meta-acrylate (PMMA), polyvinyl chloride (PVC) or glass; usually 1.2 mm thick), a precoat layer, the recording medium, a reflecting layer (aluminum) and a protective coating. Not all applications require the use of all layers, as can be illustrated by considering media in which strong absorbtion is desired where the antireflection layer is availed; the precoat layer is applied in reversable media.

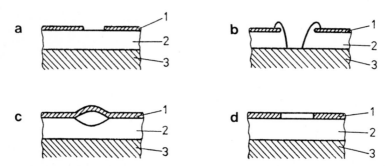

Fig. 5.1a–d. Multilayer film for real time optical recording: (*1*) recording layer; (*2*) protective layer; (*3*) substrate (glass, polymer). (**a**) Flat hole in the recording layer. (**b**) Deep hole in the recording and protective layers. (**c**) Light-induced bubble. (**d**) Reversible recording (light-induced changes of optical constants)

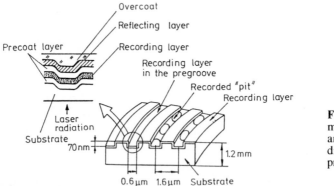

Fig. 5.2. A typical multilayer structure of an optical memory disc with tracking pregroove [5.5]

On the surface of the substrate, spiral (or centroidal) grooves are precut for the photoelectrical tracking servomechanism. Typically, the dimensions of the tracking pregrooves are 1.6 µm, 0.6 µm, and 70 nm, respectively (Fig. 5.2). The smallness of these dimensions demands utmost in cleanliness of the manufacturing conditions (rooms of class A-10 with not more than 10 dust particles per cubic foot) and the best in vacuum deposition equipment and

Table 5.2. Thermal properties of substrates used in optical discs [5.8]

Material	Glass transition temperature K	Thermal conductivity W/cm·K	Photothermal relaxation time for a 1 μ - diam. spot ns
Fused quartz (a-SiO$_2$)	2216	0.0138	78.0
Polymethylmetaacrylate (PMMA)	383	0.0020	890.0
Polyvinylchloride (PVC)	350	0.0017	–

computer robotics; i.e., optical disc production is not easier than microelectronic megabit chip manufacturing [5.1,2,7].

The substrate of the multilayer film not only acts as the track for the reading/writing head but also interacts with the recording medium. In the photothermal recording step, the substrate must be stable to temperatures of between 500 to 800 K, i.e., it must be chemically inert at the transition temperature of the recording medium. It must also have a low thermal conductivity to reduce thermal losses from the recording medium to the substrate. It has been shown that a small change in thermal conductivity can increase the required incident light intensity by several times, because the thermal conductivity (i.e., thermal losses) occurs over the complete recording time. Thus, the thermal losses tend to increase with increased laser pulse duration (Sect. 5.2.2). The properties of typical substrate materials are collected in Table 5.2.

The precoat layer, (also known as capping or internal protective layer) is important in reversible recording. It has several protective functions, namely, to prevent diffusion of reaction products out of the recording layer and to prevent their oxidation, or other chemical reactions with the reflection layer, or the substrate, and the atmosphere. The precoat layer must be transparent to the laser light, stable at the transition temperature of the recording medium, and have low thermal conductivity. The chemical composition of this layer depends on the recording medium, but, generally, oxide compounds such as SiO_x, ZnS, or organics with low thermal conductivity, are used.

The reflecting layer is usually an aluminum film with a thickness of 50 nm, or more. Having a reflecting layer improves both the recording conditions, since, now the laser light crosses the recording medium twice, and the readout setup (optical head) has a simpler construction. Finally, the protective coating seals the entire multilayer structure. It is generally composed of SiO_x or organic laquers (acrylics, cellulose acetates) in layers of several hundred microns.

5.1.2 Components of the Recording and Reading System

Figures 5.3 and 5.4 show the components of an optical disc recording and reading system. Light from a diode laser is focussed by a series of lenses to a spot of ≤ 1 µm diameter, split into recording (or readout) and control beams onto the optical disc and the photodetectors. For reversible discs, the same optical system is used for recording and for readout (the recording takes place on an area of ~ 1 µm^2, Fig. 5.4). The optical recording system consists of three main components: the optical disc (OD), the arm and the electromechanical drive system (Fig. 5.3). The optical disc has been discussed above. The arm is an electromechanical system with an optical "head" which allows the light beam to follow the information track and to focus exactly onto the correct position. The electromechanical system ensures that the disc is always rotating at the prerequisite speed.

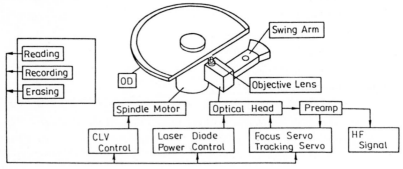

Fig. 5.3. Control system for disc tracking and optical processing with a constant linear velocity (CLV) from [5.5]

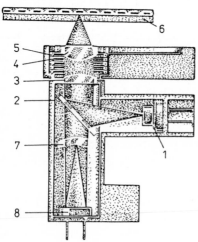

Fig. 5.4. A simple optical head construction ("light pen") for memory discs: (*1*) diode laser; (*2*) half-silvered mirror; (*3*) collimator lens; (*4*) objective lens for focusing the light beam; (*5*) actuator for positioning the objective; (*6*) optical disc; (*7*) beam splitter; (*8*) photodetector [5.10]

A simple optical head is shown in more detail in (Fig. 5.4). It has several components: (1) a laser diode, (2) half-silvered mirror, (3) collimator lens, to make the laser diode beam parallel, (4) objective lens, to focus the light beam onto the recording medium, (5) focus actuator, to control the position of the objective lens relative to the disc (6), (7) beam-splitter, (8) photodetector (in many optical head systems more than one photodetector is used). Generally, diode lasers are used with an emission between 780-850 nm (Chap. 3). The laser beam is first collimated, and, then focused through a transparent substrate (Fig. 5.2) or a transparent overcoat (if the illumination is from the opposite direction). After reflection from the reflecting layer, the light is collected by the same objective and directed to a photodetector.

The optical system produces a small spot of diameter $2\,r \approx \lambda = 0.8\ \mu m$ (the depth of the foussed light is typically about 1.5 μm, for most optical read-out systems). The illumination at recording (or readout) usually takes place from the substrate side (standard thickness is 1.2 mm). Thus, the diameter of the laser beam on the substrate is much larger than the spot, so that any damage on the substrate surface will have, practically, no effect on the signal to be read out. The optical disc has no contact with the optical head (they are sparated by about 1 mm) and, therefore, no wear occurs.

The tracking of the disc is guided by the pregroove structure on the disc (Fig. 5.2) and a multiple photodetector system (for correct tracking and focus depth control, up to 6 photodiodes are used). Holographic optical elements instead of lenses, are common (they are cheaper and are lighter in weight). However, the tracking can also be controlled with a simpler system (Fig. 5.4) by measuring the light intensity of the reflected recording or readout beam [5.10]. For an optical phase recording, the change in the reflectivity on the recorded spot is usually between 20 and 30%. After amplification, this gives an SNR of between 40 and 50 dB.

For magneto-optic materials, the recording process and readout are more complicated: recording occurs in a magnetic field and readout (via the Faraday effect, Chap. 3), a more sophisicated optical head, with a polarization device, must be used (Sect. 5.2.4).

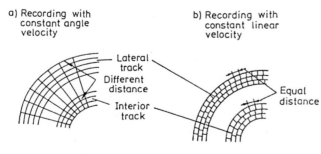

a) Recording with constant angle velocity

b) Recording with constant linear velocity

Lateral track

Different distance

Interior track

Equal distance

Fig. 5.5a, b. Different recording modes in optical memory discs. (a) recording with constant angular velocity. (b) recording with constant linear velocity (25% higher total storage capacity)

Tracking, can be at constant linear, or constant, angular velocity, depending on the distribution of data on the disc (Fig. 5.5). Discs read at constant linear velocity allow about 25% more storage capacity.

5.2 Photothermal Recording

As mentioned above, there are two classes of commonly used recording materials: phase-change and magneto-optic. Phase-change materials have now been studied for more than two decades. Investigations began with the work of *Ovshinsky* et al. [5.14], while early works dealt mostly with tellurium based alloys [5.15–17] which we shall discuss in detail.

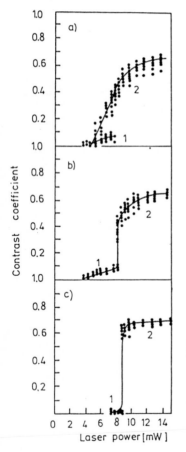

Fig. 5.6a–c. Threshold power for optical recording in Te and Te-alloy films, 30 nm thick on a PMMA substrate; excitation with a Kr^+-laser at 647 nm; pulse length of 75 ns: (**a**) Te; (**b**) TeBi; (**c**) TeGe; (*1*) reversible phase transition; (*2*) open hole formation (ablative recording). From [5.18]

Photothermal recording is a process in which the absorbed light energy is transformed to phonons and the medium is heated to a temperature of phase transition (glass transition temperature T_g, melting point T_m, evaporation point T_{ev}). Reversible recordings can be made when the crystal-to-glass transition point is reached; nonreversible ones, when holes are created by evaporating atoms. Therefore, the photothermal recording takes place at a threshold light intensity, I_{th}, which corresponds to the temperature of phase transition. The threshold laser power required to induce phase transitions in Te, TeBi and TeGe films, are shown in Fig. 5.6 [5.18]. The threshold power vary for different compounds and is lowest, 4 mW, for tellurium films (for a spot diameter of 1 μm, this corresponds to the threshold light intensity $I_{th} = 5 \cdot 10^5$ W/cm^2). Of course, reversible phase transitions occur at lower laser powers, while open hole formation requires higher intensities ≥ 8 mW. The presence of the optical recording was measured by the contrast coefficient, k_γ, equal to:

$$k_\gamma = \frac{I_1 - I_2}{I_1 + I_2} , \tag{5.1}$$

where I_1, I_2 are the transmitted light intensity in the recorded region, and outside it. Reversible recording requires lower powers; however, Fig. 5.6 shows that their contrast coefficient is small, $k_\gamma \approx 0.05$. Open hole formation yields images with higher contrast. At a saturation power of $\sim 14-16$ mW, the contrast coefficient is: $k_\gamma \approx 0.65$ (Fig. 5.6a).

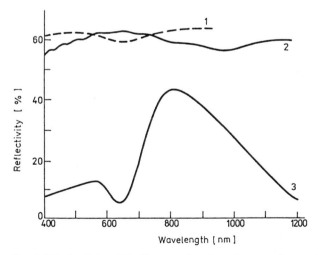

Fig. 5.7. Reflectivity of Te films as a function of evaporation rate in the presence of an anti-reflection layer: (*1*) rapid evaporation; (*2*) slow evaporation; (*3*) Te with an anti-reflection TeSe coating. The reflectivity is minimized for He-Ne laser emission 632.8 nm. [5.20]

The optical properties of Te-films depend on the rate of evaporation and on whether an antireflection coating is present (Fig. 5.7). The contrast coefficient, k_γ, of Te films also depends on the overcoat thickness (Fig. 5.8). At higher thicknesses, k_γ decreases.

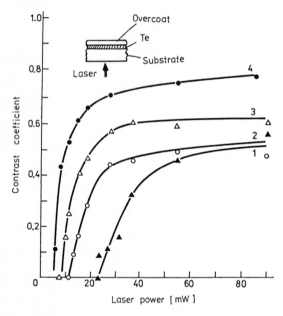

Fig. 5.8. Contrast coefficient vs. laser power (measured at the focusing lens) for 30 nm thick Te films on PMMA substrates coated with several thicknesses of SiO$_2$ layers: 1 - 5000 Å; 2 - 1000 Å; 3 - 300 Å; 4 - without SiO$_2$. The ablative writing experiments were performed using 200 ns Kr$^+$-laser pulses at 647.1 nm. [5.20]

5.2.1 Irreversible Recording in Tellurium Compounds

Tellurium compounds were used early in irreversible recordings by open hole formation [5.18, 19]. The efficiency of the process depends on the laser heating temperature (T_0 in the center of the illuminated region), the surface tension coefficient, α_s, thickness, d, and viscosity, η_v. The relaxation time, τ_{rel}, i.e., the time required for the hole edges to freeze, is given by:

$$\tau_{rel} = \frac{r^2 \eta_v}{3 d T_0} \cdot \left(\frac{\partial \alpha_s}{\partial T}\right)^{-1}, \tag{5.2}$$

where α_s is the surface tension coefficient and T is the temperature. The estimated value of τ_{rel} limits the length of the laser pulse that can be used

Table 5.3. Some optical recording characteristics of tellurium

Viscosity in kg/(m·s)	0.001
Surface tension coefficient α_s in J/m^2	0.178
Surface tension temperature coefficient $\partial\alpha_s/\partial T$ in J/(m^2 K)	$24\cdot10^{-6}$
Melting point in K	725
Density in kg/m^3	$6.3\cdot10^3$
Thermal conductivity in W/(m·K)	3.2

for ablative recording, $\tau_{las} \geq \tau_{rel}$. Optimally, $\tau_{las} \approx \tau_{rel}$, otherwise, the recording medium is slow.

The thermal parameters of Te are presented in Table 5.3. The open hole relaxation time for a laser spot of 1 μm diameter, according to (5.2), gives: τ_{rel} ~ 100 ns, in good agreement with experiment [5.18]. The recording parameters of Te films (Table 5.4) are typical for chalcogenide recording materials. Early studies of photothermal recording dwelled on elementary semiconductor and metallic systems (Se, Te, Bi) and their alloys. For binary, and more complicated media, the thermal properties, in particular melting points, must not be very different, because of thermal decomposition during preparation and optical recording. Thermal parameters of other chemical elements are shown in Table 5.5, which might be useful as components for optical storage media.

Table 5.4. Recording parameters of tellurium films [5.21]

Recording parameters μm)	Lasers	
	10mW He-Ne or diode laser (0.8 μm)	100 mW Ar$^+$-laser or 20 mW diode laser (0.8
Recording rate in Mbit/s	10	>50
Number of bits on a 30-cm dia. disc	$7\cdot10^{10}$	$2\cdot10^{11}$
Linear recording density in μm/bit	1	0.5
Signal-to-noise ratio in dB	40	40
Recording energy in nJ/bit	2	2
Storage time in years	10	10

Table 5.5. Thermal and recording parameters of materials for optical memory discs

Material	Melting point K	Boiling point K	Specific heat of melting J/g	Specific heat of evapor. J/g	Thermal conduvtivity at T = 300 K W/(cm·K)	Specific heat at T=300 K J/(g·K)	Threshold power[a] mW
Al	932	2740	398	10852	2.37	0.900	15
Ti	1933	3550	324	8828	0.219	0.523	12
Te (rhomb.)	725	1260	140	894	∥ c 0.0338 ⊥ c 0.0197	0.201	3
Se (monocl.)	490	958	66	800	∥ c 0.0452 ⊥ c 0.0131	0.321	–
Se (amorph.)	–	–	–	–	0.0052	–	–
S (rhomb.)	386	718	54	2836	0.0027	0.733	–
As (grey)	1090 at 28 atm	–	–	370	0.502	0.330	–
Sb (trig.)	904	2023	165	1050	0.244	0.205	–
Bi (rhomb.)	544	1833	52	725	∥ c 0.0530 ⊥ c 0.0919	0.125	3
Pb	600	2013	24	867	0.353	0.128	3
Ge	1210	3103	470	4600	0.602	0.322	9
In	430	2353	28	1970	0.818	0.234	5

[a] For a spot with a diameter of 1 μm; the SNR for a recorded bit is of the order of magnitude of between 40 and 50 dB

5.2.2 Reversible Recording in Tellurium Compounds

The key attributes of good reversible optical recording media are fast writing and erasing, adequate number of reversible cycles, stable written and erased states, good signal-to-noise, and high light, sensitivity.

Rapid progress in developing reversible media began in 1983 with a report by *Clemens,* [5.22] who demonstrated that as much as 40 000 reversible cycles between the amorphous and crystalline states could be accomplished. In the same year, *Takenaga* et al. claimed to have seen 10^6 cycles on a disc with 55 dB signal-to-noise ratio (SNR) using a tellurium oxide based active layer [5.23]. The recording process of tellurium oxide based materials was shown to take up to several minutes to complete [5.24]. This is a potential problem for many reversible media; in addition, the minimum erasure time of 1 μs is too long for most applications.

In 1986 *Chen* et al. reported that erasure times of less than 30 ns could be achieved in phase-change recording without compromising data stability by using media based on non-stoichiometric compounds [5.25]. This also allowed the use of a much simpler optical head consisting of a single laser beam.

Other significant developments, reported in 1989, include those by *Sato* et al. [5.26], who reported 10^5 reversible cycles using In-Sb-Te; *Terau* et al. obtained more than 10^5 cycles in Ge-Te-Sb-Co films [5.27]; *Ochta* et al. observed $2 \cdot 10^6$ reversible cycles in Ge-Te-Sb-based recording layers, using the most stringent criterion, namely, bit error rate, to monitor failure [5.28]. *Rubin* achieved 10^7 reversible recording cycles in Te films using the phase transition between the amorophous and the crystalline states [5.4].

In phase-change recording materials, with a transition from an amorphous to a crystalline state, the transmission change is usually of the magnitude of between 20 to 30% (irradiation by a diode laser emitting in the infrared between 0.78 and 0.86 μm at 10–20 mW and with a beam cross section of 1–2 μm^2). Recording, erasing and readout are all done with the same laser but at different powers.

Therefore, the wavelengths of the absorption before and after the phase change must be near the emission line of the laser. Since the storage density augments with decreasing wavelength: $N_{inf} \sim 1/\lambda^2$ (Chap. 2), these media will become even more attractive when shorter wavelength lasing diodes become available.

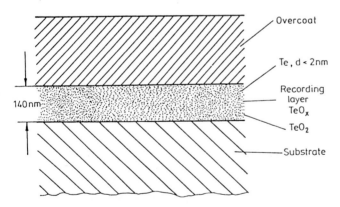

Fig. 5.9. Proposed structure of TeO$_x$ films. [5.23]

TeO$_x$ Films. The structure of the TeO$_x$ film is heterogeneous (Fig. 5.9), the upper part of the TeO$_x$ layer containing a thin (2 nm thick) Te layer. The capping layer can be TeO$_2$ or SiO$_2$. These non-stoichiometric TeO$_2$ - Te (TeO$_x$), and impurity doped films, show yet another type of reaction which can be used for optical recording purposes. The Te or other impurity atoms (Ge, Sn, In) in these films form colloids which, under a long light pulse of 0.1–10 μs coagulate, and with shorter pulses of 0.05–2 μs, dissipate:

$$\text{TeO}_2 + \begin{array}{c}\text{small}\\\text{particles}\end{array} \underset{\overset{\textit{short puls}}{\longleftarrow}}{\overset{\overset{\textit{long puls}}{\longrightarrow}}{}} \text{TeO}_2 + \begin{array}{c}\text{large}\\\text{particles}\end{array} \qquad (5.3)$$

A model of the reaction (5.3) was proposed by *Tyan* et al. and is shown in Fig. 5.10, [5.29]. The initial as-deposited film of TeO_x contains tiny grains of crystalline Te dispersed in a glassy TeO_2 matrix (Fig. 5.10a). There are two different processes: (1) direct laser-induced coagulation of the grains (Fig. 5.10b) or (2) thermally induced grain coagulation followed by a laser writing pulse (Fig. 5.10a – c). Optical erasure from the final state (Fig. 5.10c) is effected by irradiation with a short laser pulse which destroys the large particles. A large number (approximately 10^6) of reversible recording-erasure cycles can be attained. However, the application of this TeO_x system is limited by the presence of slow dark reactions in the coagulated Te grains (Fig. 5.10c). The coagulated amorphous Te particles are not stable and crystallize spontaneously, with a relaxation time from several seconds to several minutes.

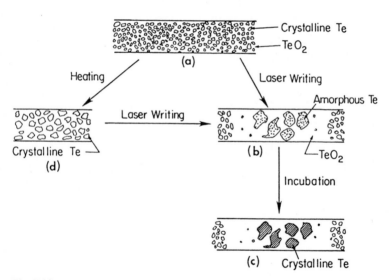

Fig. 5.10. Proposed mechanism of optical recording in TeO_x films. [5.29]

The TeO_x system demonstrates the dependence of the recording parameters on the stoichiometric composition (Fig. 5.11). A stable optical recording, capable of long-time information storage, is possible only at $x = 1.2$. The stoichiometric composition also determines the phase transition temperatures, of course, and thus, the necessary laser powers. For $x = 1.2$ the phase transition temperature, measured by thermal annealing, is 130° C (Fig. 5.12,3).

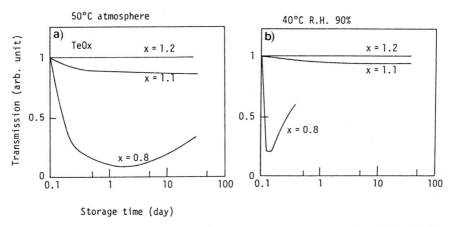

Fig. 5.11a, b. Transmission change after optical recording in TeO$_x$ at T = 323 K. (1) x = 0.8; (2) x = 1.1; (3) x = 1.2. [5.23]

As has been mentioned, in practice, the recording, readout and erasure must be possible with one and the same diode laser, emitting in the spectral region 0.78–0.86 µm. This means that the photoinduced absorption or reflectivity changes of the medium must be large enough and in the spectral region not far from the recording wavelength. Figure 5.13 shows that in TeO$_2$ this change is about 20%, which gives a high signal-to-noise ratio (SNR) of 59 dB [5.30].

Similar recording processes with the same parameters also takes place in nonstoichiometric SbO$_x$ [5.31].

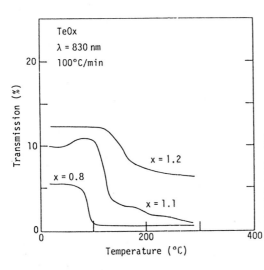

Fig. 5.12. Transition temperature of TeO$_x$ films with various compositions, x: (1) x = 0.8; (2) x = 1.1; (3) x = 1.2 (wavelength λ = 830 nm; heating rate 100 K/min). The transition temperature corresponds to that region of rapid decrease of transmission. From [5.23]

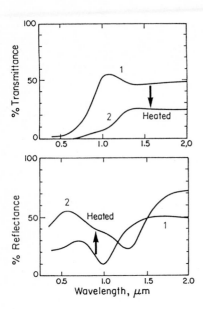

Fig. 5.13. Transmission (**a**) and reflectance (**b**) change in TeO_x films before (*1*) and after (*2*) heat treatment (T = 620 K, several minutes). [5.29]

$(GeTe)_xSn_{1-x}$ Films. As might be expected, the crystallization and amorphization times of the optically active layer are dependent on the composition, which in turn is a function of the thermal treatment [5.4]. Minimizing the crystallization time is important for achieving rapid optical erasure. The best time, so far, was found for $(GeTe)_{0.85}Sn_{0.15}$. The amorphization time is usually determined from the cooling rate of the melted state. The crystallization time is shorter for laser-amorphized samples (200 ns, cooling rate $10^9 - 10^{10}$ K/s) than for as-evaporated samples (700 ns), which are more rapidly quenched (with the rate $10^{12} - 10^{14}$ K/s). However, the laser-amorphized samples contain some residual crystal nuclei in the irradiated spot. The crystallization of as-evaporated amorphous $(GeTe)_{0.85}Sn_{0.15}$ films was observed as a function of laser power and pulse length (Kr^+-ion laser emission 647 nm) for a spot with a diameter of 1 µm. Three different power-time regions were found. (1) For pulse lengths of $\tau \approx 300$ ns and a laser power $P < 9$ mW, no crystallization was observed. At higher laser power, only initiating micro-crystallites were observed for 300 ns pulses. (2) For $\tau \approx 500$ ns and $P = 7.4$ mW, half-complete crystallization was observed, and, (3) only for $\tau \approx 700$ ns, a complete crystallization of the spot was observed. While the as-evaporated samples have a longer crystallization time than laser-amorphized spots, for memory systems, only laser amorphization would be required.

5.2.3 Photothermal Recording in Antimony Compounds

These compounds are attractive for optical recording, primarily because of the short crystallization time (approximately 30 ns), as observed by *Barton* et al. in Sb_2Se and $GaSb$ films [5.32]. The main result from [5.32] was that the crystallization time for chemical compounds is shorter than for alloys. This must be taken into account in further research into optical materials.

Sb_xSe_{1-x} Compounds. Reversible amorphous $\Leftrightarrow$ crystalline phase changes in films of Sb_xSe_{1-x} ($0.4 \leq x \leq 0.7$) under pulsed laser excitation have been studied [5.33] by optical (absorption spectrum, measurement of the birefringence in the crystalline phase) and X-ray diffraction studies. In these compounds the reactions are also photothermal, requiring a threshold laser intensity to proceed. The kinetics have been analyzed in a wide range of light intensities up to 10^6 W/cm^2 and pulse lengths $\tau \geq 100$ ns. In the thermal amorphous-to-crystalline transition, two metastable crystalline phases of Sb_2Se were found, one of them had been previously observed in [5.32].

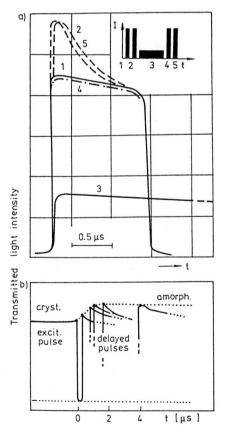

Fig. 5.14a, b. Reversible photo-induced phase transitions between the crystalline (*c*) and the amorphous (*a*) states in $Sb_{0.65}Se_{0.35}$ films at He-Ne laser excitation. (**a**) 1,4 - photoamorphization (short pulses, c $\rightarrow$ a); 3 - photocrystallization (long pulse, a $\rightarrow$ c); 2, 5 - photoinduced reaction (short pulse, a $\rightarrow$ a). 1, 2, 4, 5 - pulse length 1.25 μs, power 12.4 mW; 3 - pulse length 4 μs, power 3.3 mW. (**b**) Amorphization kinetics measured by delayed pulse probe transmission (pulse length 1 μs). The rate of photoamorphization was measured by the amount of transmitted light as a function of time. Amorphization takes place at $\tau_a \approx 1$ μs

Optimum recording parameters were achieved in a polycrystalline $Sb_{0.65}Se_{0.35}$ film (thickness d ≤ 100 nm) with a ZnS underlayer and ZnS overcoat. The amorphization takes place after irradiation by a short pulse, $\tau \leq 1$ µs (Fig. 5.14a, 1). A laser pulse with intensity higher than the threshold intensity ($I \geq I_{th} \approx 10^5$ W/cm^2) heats the active layer up to the melting point. Amorphization occurs during rapid cooling. The cooling rate was theoretically estimated and experimentally measured from the optical transmission kinetics of the laser pulse (Fig. 5.14b) and is approximately 10^9 K/s.

A second short pulse (Fig.5.14a, 2) does not change the ultimate amorphous state (reaction a → a). Photocrystallization (a → c) takes place during the longer laser pulse of lower intensity (Fig. 5.14a, 3). Further excitation with short pulses (Fig. 5.14a, 4 and 5) repeats the behavior of the processes under pulses 1 and 2. The rate of amorphization was measured by observing the transmission of a sequence of probe pulses with different delay times (Fig. 5.14b) during the cooling of the melted films. Amorphization occurs in less than $\Delta t \leq 1$ µs.

The photocrystallization is a complex process, which depends on the temperature, atomic diffusion rate and crystallization energy (released during crystallization). The photocrystallization can take place in the liquid or in the solid state. Usually, in optical information recording in thin films, solid state photocrystallization is used with a phase transition temperature of $T_{ph} \approx 0.5\, T_m$ (T_m is the melting point of the bulk material) [5.4,13]. The photocrystallization rate for a given compound depends also on the composition [5.34,35]. The shortest crystallization time, ~ 30 ns, was found for $Sb_{0.63}Se_{0.37}$ with 10^5 reversible a ⇔ c cycles [5.25].

A model of photothermal reactions in thin films has been proposed for both cw and pulsed laser excitation [5.33]. According to this model, the rise of the temperature during the pulse is:

$$\Delta T(t) = It / \left(c_0\rho_0 d + 0.5 \cdot c\rho\sqrt{\pi a_T t}\right),\tag{5.4}$$

where c_0, ρ_0 and c, ρ are the heat capacity and density of the film and the substrate, d is the thickness of the film and a_T the temperature conductivity of the substrate.[1] The threshold energy for photocrystallization in the solid state depends on the pulse length, τ. At larger t ($t \to \tau$), when the second term in (5.4) exceeds the first, the optically induced heating is propotional to the squareroot of the pulselength:

$$\Delta T(t) \sim I\sqrt{\tau} = \text{const} .\tag{5.5}$$

---

[1] For CW excitation the light induced temperature increase $\Delta T = T - T_0$ (where T_0 is the initial temperature) is equal to $\Delta T = I \cdot r / K$, where r is the radius of the illuminated area and K is thermal conductivity of the substrate (Chap. 4, Fig. 4.20)

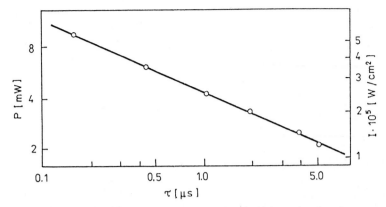

Fig. 5.15. The threshold laser power and threshold intensity for photocrystallization of $Sb_{0.65}Se_{0.25}In_{0.1}$ films (thickness $d = 80$ nm) vs. laser pulse length τ. The spot diameter is $2\,\mu m$

The experimentally observed dependence in $Sb_{0.65}Se_{0.35}In_{0.10}$ is in agreement with model (5.5) and demonstrates the thermal origin of the process (Fig. 5.15).

SbIn and SbSeIn. $Sb_{0.45}Se_{0.2}In_{0.35}$ and $Sb_{0.6}In_{0.4}$ films were developed in the Fujitsu Research center and since 1988 have been produced as reversible optical memory discs [5.36]. These discs were produced by separate Sb, Se and In evaporation in a multilayer system with a 100 nm thick active recording layer and ZnS underlayers (similar to Fig. 5.2).

The optical recording-erasure processes in SbSeIn and SbIn systems are different. In the $Sb_{0.45}Se_{0.20}In_{0.35}$ system, a reversible amorphous $\Leftrightarrow$ crysralline phase transition takes place. In $Sb_{0.6}In_{0.4}$ films, the phase-change reaction is more complicate than in Sb_xSe_{1-x}, involving two crystal phases with different melting points which was studied by electron microscopy, microdiffraction and microanalysis methods. As usual, the recording step requires a short laser pulse with a high intensity. This melts the irradiated spot and causes Sb atoms to diffuse from the outer part (the solid phase) towards the center (the liquid phase, Fig. 5.16a). This process results in the formation of large crystalline Sb grains in the center of the spot and SbIn along the edges. The erasure of $Sb_{0.6}In_{0.4}$ films require both stronger and longer laser pulses than used in writing. This lowers the mobility of Sb atoms, therefore, smaller Sb crystals are created (Fig. 5.16b). Thus, the phase composition of $Sb_{0.6}In_{0.4}$ films is different after writing and erasing. The reversible reaction is $Sb_{0.6}In_{0.4}$ $\Leftrightarrow Sb_{0.5}In_{0.5}+Sb$; because thermodynamically, with respect to crystallization of amorphous materials, such a crystal 1 $\Leftrightarrow$ crystal 2 phase transition is better for long time information storage.

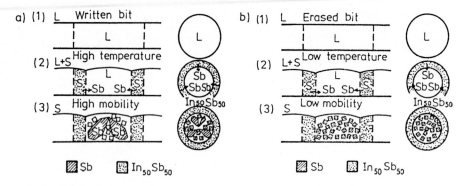

Fig. 5.16a,b. Photoinduced phase transition in $Sb_{0.6}In_{0.4}$ films with diode laser ($\lambda = 830$ nm) excitation. Film thickness: $d = 100$ nm; spot diameter: 1 μm. **(a)** Recording power: 8 mW, pulse length 170 ns. **(b)** Erasure power 3 mW, pulse length 500 ns. (*1*) laser-induced melting; (*2*) diffusion from the solid to the liquid state during cooling; (*3*) final solid state. L is the liquid state and S is the solid state. [5.36]

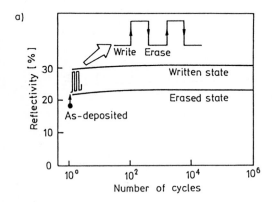

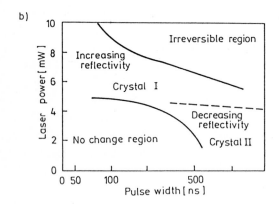

Fig. 5.17 (a) Photoinduced reflectivity changes in $Sb_{0.6}In_{0.4}$ films. **(b)** Reversible and non-reversible photoinduced reactions as a function of the laser power and pulse width. From [5.36]

The optimum laser power for recording was approximately 8 mW, with 200 ns pulses, and for erasure: − 3 mW and 500 ns pulses. At higher laser powers, P > 8 mW and $\tau \geq 100$ ns, nonreversible hole formation was observed. The primary reflectivity change of 10% (Fig. 5.17a) was sufficient for a final SNR of 45 dB.

In Fig. 5.17b, the laser threshold power via pulse length for the optical recording in $Sb_{0.6}In_{0.4}$, is shown. There are three different power-pulse length regions. No phase transitions take place at short pulses and low laser power and for long higher powered laser pulses the reaction becomes nonreversible (i.e., open hole formation). Reversible optical recording is only possible in the intermediate region. Such conformation is typical for any photothermal phase change material.

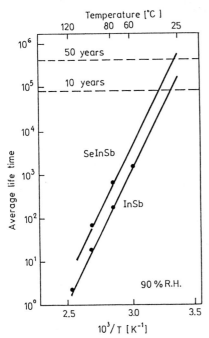

Fig. 5.18. Arrhenius plot of medium lifetime versus aging temperature. The maximum lifetime was defined as that time at which the signal to noise ratio was reduced by 3 dB, from the initial value. [5.36]

The storage time for $Sb_{0.6}In_{0.4}$ was estimated to be more than 10 years (Fig. 5.18). This result was obtained from accelerated thermal aging measurements.

Optical recording in SbSeIn films system has approximately the same recording parameter as $Sb_{0.6}In_{0.4}$. However, the storage time in SbSeIn films is better (Fig. 5.18) [5.36].

The Photothermal Recording Energy. The light sensitivity for photothermal recording is usually defined as S_{dig}^{-1}, i.e., as the exposure value for one recorded bit (Sect. 2.5). The integral light sensitivity, S^{-1} (J/cm^2), is then related to S_{dig}^{-1} over the data density (N [bit/cm^2]), i.e., S^{-1} [J/cm^2] $= N \cdot S_{dig}^{-1}$. However, for the recording process, the characteristic value of the absorbed energy is necessary. The recording energy, ΔE_{rec}, is approximately equal to the exposure for one bit (S_{dig}^{-1}) in the elemental volume of the absorbed light energy. Thus, the smallest recording area for one bit is approximately a spot with the size of the wavelength (Chap. 2, equation (2.22)) and the recording film thickness is of the order of d from 60 to 70 nm, we can estimate the elemental volume for recording one bit by: $\Delta V \approx \pi r^2 \cdot d \approx 0.03$ µm^3 (for $\lambda \approx 2\ r = 800$ nm and $d \approx 65$ nm).

The photothermal recording process, as mentioned above, is generally a light induced local heating of the recording medium, with an ensuing phase transition (amorphization, crystallization, evaporation). Therefore, we can estimate the upper limit of the reversible photothermal recording as the melting energy of the illuminated spot, and the nonreversible photothermal recording (e.g., open hole formation) as the evaporation energy for the volume of one recorded bit. For such, simple estimation, we used the thermal properties of Tellurium (Table 5.5)

Table 5.6. Optical recording energy in chalcogenide films

Recording process	Material	Recording energy ΔE [nJ/bit]	
		Calculated	Measured
Evaporating	Te	0.210	1.3
Melting	Te	0.043	---
Photoinduced phase transition	TeSbSe	---	0.4
	Sb$_{0.65}$S$_{0.35}$	---	1.6

for the volume element of the recorded spot as $\Delta V \approx 0.032$ µm^3. The results of such an estimation are represented in Table 5.6. The nonreversible recording energy (evaporation) for one bit in Te films is only five times larger than for reversible recording (melting). However, the experimental value of the nonreversible recording is remarkably higher than the calculated energy which demonstrates the high thermal energy losses in the substrate (these losses were not taken into account in the estimation). The light sensitivity of different phase change recording materials (Table 5.6 and 5.7) is of the order of 0.4 to 10 nJ/bit, which is higher than the evaporation energy for one recorded bit in Te films. This demonstrates the low recording efficiency of this class of recording materials. Such recording efficiency leads to a strong electron-phonon interaction in the recording material and is typical for this class of recording materials. However, the recording energy and the recording costs in common magnetic discs or tapes are much more expensive than in phase change and magneto-optical materials. The low recording costs of optical memory discs (in spite of the low photothermal efficiency) is from one to two orders smaller than for magnetical discs or tape. This has stimulated the wide field of application optical memory systems. New optical phase change materials have been studied, and the parameters of these new materials are shown in Table 5.7.

Table 5.7. Recording characteristics of erasable phase change materials (amorphous $\Leftrightarrow$ crystalline)

Material	Light sensitivity Recording nJ/bit	Erasure nJ/bit	Minimum recording time μs	Over-writing cycles
GeSbTeCo ($Ge_1Sb_2Te_4$+Co) [5.38]	2.4	4.5	0.4	10^4
$Ge_{12}Sb_{39}Te_{49}$ [5.39]	1.5	10	0.1	$>10^2$
GeTe-Sb_2Te_3-Sb [5.40]	2.0	10	0.2	10^5
Ge-Te-Sb [5.41]	4.0	5.0	0.3	10^6

Reversible phase change optical discs will be more widely produced in the near future [5.42]. However, the recording time is limited by the heating rate of the recorded spot (for 1 μm size, approximately 10 ns) as well as the light sensitivity (1 – 10 nJ/bit, Table 5.6 and Table 5.7).

5.2.4 Magneto-Optic Recording

Magneto-optic recording is also a photothermal process which can occur in ferroelectric or ferrielectric materials. The optical recording is conducted in an external magnetic field of several hundred Oersted ($\sim 10^4$ A/m). The magnetic field and the light spot are concentrated onto a small area, $1\mu m^2$. Generally, permanent samarium type magnets or solenoids are used [5.43–45]. The laser pulse induces photothermal heating to the magnetic phase transition temperature. The medium is then rapidly cooled to fix (freeze) the magnetization change.

There are at least two different magneto-optical recording processes in the medium: (1) photothermal heating of a ferromagnetic to the Curie point, T_c, with a subsequent local change to the paramagnetic state, and (2) photothermal heating of a ferrimagnetic material to the compensation temperature, T_{comp}, at which one of the magnetic sublattices is reoriented [5.44–47]. The second process is more convenient, and, thus, more common for reversible optical recording because the compensation temperature is lower than the Curie point.

The following conditions are desirable for magneto-optic recording materials: stable written and erased states, that is, stable magnetic sublattices for different magnetization states; suitable compensation temperatures, i.e., T_{comp} must be high enough for longtime information storage (approximately 10 years) but low enough for high light sensitivity. Usually, T_{comp} must be in the region of 400–450 K.

Table 5.8. Characteristics of magneto-optic recording materials

Material	Thickness nm	Phase transition-point K	Coercive force, H_c kOe	Recording energy nJ/bit	References
Tb-Fe	100	413	2	5	[5.48]
$Tb_{0.21}Fe_{0.79}$	40	400	1.5	0.5	[5.49]
GdTbFe	50	430	0.6	0.5	[5.50, 51]
GdTbCo	50	420	2.0	0.5	[5.50]
$Gd_{0.24}Fe_{0.76}$	70	300	0.5	0.3	[5.51]

In recent years there has been progress made in magneto-optical recording in rare earth magnetic materials (e.g., GdTbCo, TbCo, TbFe, TbFeCo). These materials have a compensation temperature in the region of 400–450 K and are being developed by many companies (Philips, IBM, Fujitsu, Hitachi and many others). Characteristics of common magneto-optic media are given in table 5.8.

In Fig. 5.19, the reorientation of the direction of magnetization, in the magnetic layer (GdTbCo or other) during recording (laser pulse in an external magnetic field) and erasure (laser pulse without a magnetic field), is illustrated. To restore the system to zero magnetization, a coercive force is needed. At room temperature, for GdTbCo, this force corresponds to: $H_c \approx 2$ kOe, while at the compensation temperature of 420 K H_c, drops below 200 Oe. This can easily be delivered by a permanent magnet, or solenoid, under light induced warming. Diode lasers, with light emission of between 780 and 850 nm, are used in

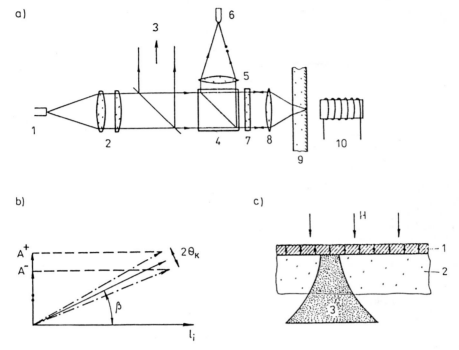

Fig. 5.19a-c. A magneto-optic switching device and magneto-optic recording in ferrimagnetic garnet films. (**a**) Recording device: (*1*) diode laser; (*2*) collimator; (*3*) reflected beam for tracking and focussing; (*4*) polarized beam splitter; (*5*, *6*) focussing lens and photodiode for readout; (*7*) Faraday rotator (β is the angle of the total rotation of the polarization plane); (*8*) focussing lens; (*9*) recording medium; (*10*) coil for local magnetic field generation. (**b**) Readout process leads to a change of the polarization plane $\pm\theta_K$ caused by the magnetic Kerr effect in the medium (A^+, A^- are the amplitudes of polarized light). (**c**) Recording process is a light induced local heating in an external magnetic field which reorientated the magnetization of the medium. From [5.51]

magneto-optic recordings. Some of the light reflected by the disc (Fig. 5.19a, 3) is used for positioning and focusing the laser beam on to the disc, with the tracking pregroove (Fig. 5.2).

The recording process in the medium (optical heating in an external magnetic field) leads to a local change of the magnetization of the recording medium (the direction of the external field, **H**, is antiparallel to the direction of magnetization). This causes a Faraday rotation of θ_k (the magnetic Kerr effect, Sect. 3.4.2) during readout. Thus, the light beam passes the recording medium twice (reflection from the metallic reflection layer, Fig 5.2): hence, the total Faraday rotation angle is $2\theta_k$ (Fig 5.19c). For magneto-optic recording systems usually a polarization beam splitter is necessary which splits the light into two components with parallel and perpendicular polarizations (Fig. 5.19a, 4). The parallel component is transmitted and used for tracking and focusing; the perpendicular component, for recording and readout.

Fig. 5.20. Magneto-optical recording patterns in GdTbFe film on a pregrown substrate (track spacing ≈ 1.7 µm; the length of the recorded spots ≈ 3 µm) in a polarization microscope. [5.51]

The erasure of the recorded information takes place through optical heating in an external magnetiic field, with the direction of the external field oriented parallel to the initial direction of magnetization of the medium (i.e., oposite direction than for recording).

Diode lasers with the average power of 15 mW are necessary, for optical recording and erasure, while during readout, the laser power is 10 times lower. The recording and erasure times for rare earth magnetic materials are of the order of 100 ns and the light sensitivity is approximately 0.5 nJ/bit (for a spot diameter of 1 µm). The required magneto-optical recording energy is lower than the recording energy for an optical phase-change recording medium. This is because phase transitions are generally more complicated. However, the signal-to-noise ratio attainable by magneto-optical recording materials operating on the basis of the Faraday and magnetic Kerr effect is much lower, thus, requiring a more complicated optical head (polarization analyzer) than for optical phase-change materials (Fig. 5.19a). This is in particular compensated by the use of low power diode lasers.

The magneto-optical recording pattern in a GdTbFe film is shown in Fig. 5.20 [5.51]. A detailed description of the magneto-optical storage is given by *Saffady* [5.52]. The commercial magneto-optical discs are also discussed in [5.53-55]. New Co-Pt and magneto-optical recording media are discussed in [5.56].

5.3 Light Stimulated Recombination Luminescence

Luminescence phenomena in solids are used as thermoluminescent dosimeters and radiophotoluminescent detectors [5.57]. Such luminescence arises as the result of radiation damage, i.e., radiation induced defect creation which acts as luminescent centers in photo- or thermoluminescence. The concentration of the radiation induced defects (electron trapping centers or luminescence centers) is proportional to the absorbed radiation energy.

Radiation induced luminescence can also be used for information processing [5.58-65]. Two types of reaction can be used for information recording. In the first type, illumination by light with photon energy, $h\nu_1$, induces the electron transfer reaction, i.e.:

$$A + B \xrightarrow{h\nu_1} A^+ + B^- , \tag{5.6}$$

where A is, thus, the luminescent defect center and B is the electron trapping center. If A^+ and B^- are deep acceptor and donor levels, then the material can have a long data storage time. The data can be read out by illumination with photons $h\nu_2$, which leads to a recombination luminescence reaction:

$$A^+ + B^- \xrightarrow{h\nu_2} A^* + B , \tag{5.7}$$

where A^* is an excited luminescent center which emits light quanta of $h\nu_L$, which are the readout signal. Note that this readout process also erases the initial data.

In the second type of reactions, the luminescence centers are created only by illumination with $h\nu_1$, as in reaction (5.6), where A^+ is again the light induced luminescence center. Readout is effected by irradiation of A^+ with light of energy $h\nu_2$, causing the formation of an excited state ion, yielding:

$$A^+ \xrightarrow{h\nu_2} A^{+*} \rightarrow A^+ + h\nu_L . \tag{5.8}$$

The emission from the excited state of A^+ ($h\nu_L$) is the luminescence readout signal. The readout signal In reaction (5.8) does not constitute recombination luminescence, but fluorescence, i.e., an intrinsic luminescence of the defect state A^+. The readout in (5.8) is without information losses, i.e., the number of readout cycles is not limited. For reversible information storage by reaction (5.8), optical or thermal erasure is possible.

The use of light induced luminescence in information processing has some advantages over the usual optical recording via phase transitions and optical constant changes. Firstly, the luminescence readout has a very high signal-

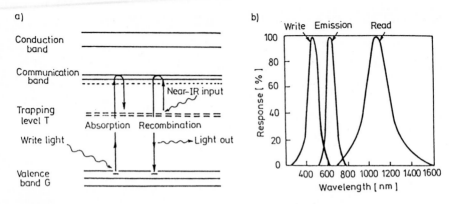

Fig. 5.21a,b. Light induced luminescence for optical recording in doped alkaline-earth metal sulphide films. (**a**) "Write light" excites the rare-earth atoms, promoting their electrons to the excited state, E, and after relaxation to the trapping levels T. Exposure to near-IR light at level T pumps the electrons back to the communication band E, from which they fall to the ground state, resulting in erasure of the data. (**b**) Writing and reading of data by the electron-trapping method of (**a**) occur at the wavelengths indicated by the respective curves. The emission curve shows the wavelengths at which the excited defects emit light upon returning to the ground state after data have been read. [5.59]

to-noise ratio. Secondly, the recording and optical erasure can be much faster (with relaxation times of nearly 1 ns) than for light induced phase transitions.

Lindmayer et al. developed a new recombination luminescence memory system in doped alkaline-earth metal sulphides [5.59-62]. The memory system was based on the recombination luminescence reaction (5.7). The principle of luminescence processing in rare-earth doped alkali-earth sulfide polycrystalline films is explained in the form of an energy level diagram in Fig. 5.21. Lindmayer used visible light ($h\nu_1 \approx 2.5$ eV) for the recording step, and, for readout (i.e., also erasure), infrared light ($h\nu_2 \approx 1.4$ eV). The wide bandgap of 4 − 5 eV of the host crystals means that the energy level of the trapping centers is quite deep, about 1 eV, giving a long storage time (ten years and more). The recording energy is about 0.2 nJ/μm^2 and the readout (erasure) energy is only 1 pJ/μm^2. Luminescence memory discs are still being developed [5.59, 60, 62, 63]. The negative features of luminescence processing in alkali-earth sulphide systems are the more complicated memory device, requiring two different lasers. A peculiarity of the recombination luminescence memory is the destructive readout, i.e., the information is erased (recombination luminescence) during the readout. This simultaneous readout-erasure process can be used in special applications as an optical processor. In classical archival memory systems, such simultaneous erasure is not permissible. However, this system is attractive for some special applications.

Special luminescent memory systems are used for X-ray imaging in medicine. The origin of these luminescent imaging systems are similar to reaction (5.7). A luminescent solid is irradiated with X-rays and electron-hole traps are created. The readout takes place by optically stimulated luminescence. *Winnacker* et al. developed a recombination luminescence system for X-ray imaging on BaFBr-Eu^{2+} polycrystalline films. The recombination luminescence reaction was stimulated by illumination of F-centers (Chap. 4) and the retrapped electrons recombine with the activator hole center (Eu^{3+}), giving the luminescence, i.e., readout signal (the luminescence emission maximum of Eu^{2+} centers is at 390 nm). For the optically stimulated recombination luminescence (readout) light in the spectral region from 490 nm to 650 nm can be used. A He-Ne laser system was used for practical application. Such luminescence memory devices can be widely used in medicine and in industry. A review of future developments is given in [5.65].

6. Summary and Looking Ahead

> Imagination is more important than knowledge.
>
> A. Einstein

Optical recording and imaging were first attempted more than 150 years ago by the inventors of silver halide photography. The progress of silver halide and color photography was long and complicated. Nowadays, photography is widely used in all branches of life – from infinite galaxies in the universum to movies, television, and hobby pictures. We have a comprehensive photoindustry producing films and cameras. However, for real time optical recording and optical memory systems conventional photography was not applicable.

Real time recording was developed much later, after the invention of lasers. Rapid progress in optical recording materials only began in the eighties. Nevertheless, the development was very fast and, today, we have an advanced industry for optical memory discs and memory systems [6.1,2]. For all that, most of these applications culminate in digital recordings for conventional computer systems. This has resulted in the present day possible potential of real time optical recording and processing being devoid of implementation.

Summarizing the review on real time optical recording materials three aspects of the problem must be brifly reviewed, namely: (*1*) present state of applications; (*2*) novel applications of common recording materials; (*3*) future research and development.

The present state of applications of optical memory systems is mostly connected with digital recording on optical discs, in addition to optical filtering for picture analysis. The future progress of these conventional applications is the use of laser diodes with shorter wavelengths resulting in an increase in storage capacity proportional to $1/\lambda^2$, (Sect. 2.4.4). Further improvement of the data rate is possible using parallel read-write systems. On optical memory discs computer generated hologram recording is also possible as first demonstrated by *Psaltis* et al. [6.3]. However, all these applications are limited by the diffraction phenomena ($d_{min} \geq \lambda \approx 1$ μm) giving, in 2-D media, a storage density of 10^8 bit/cm^2 and the photothermal relaxation time (about 100 ns for the best recording materials) limiting the data rate up to 10^8 bit/sec (for one channel).

Novel Applications of Optical Recording Materials. One further development of pertinence is (3-D) volume holographic recording resulting in considerable increase of the storage density. At present, three dimensional memory systems are being developed [6.4,5]. They have a high storage density (about 10^{12} bit/cm^3) and a parallel electronic processor for data recording (or output) which can operate at a rate of several gigabits per second. However, the value of further development of 3-D memory systems require recording materials with a higher light sensitivity and a faster response. Therefore, basic research into optical recording processes is necessary for the future progress.

New possibilities have been opened by persistant spectral hole burning (PSHB) [6.6,7]. This method in two dimensional systems allowed a storage density of 10^{11} bit/cm^2 [6.8,9]. Holographic recording and optical signal processing were demonstrated, and high speed recording and readout, within picoseconds, reached. Usually, for PSHB, low temperatures are necessary (1–4 K). However, *Arnold* et al. developed a new organic polymer recording material which allows PSHB recording and readout at room temperatures [6.10]. PSHB is a novel recording method, nevertheless, this method requires an expensive setup (dye lasers with high spectral resolution, picosecond laser pulses) which makes the applications difficult.

New possibilities in conventional recording materials have been instigated by electron beam recording and the scanning tunneling microscope (the spot size can be 1000 times smaller than those of present optical discs!) [6.11,12]. However, these recording methods are only at an initial stage of progress and will have to go a long way before they can be applied.

Future Research and Development. It is difficult to forecast the future of optical recording and information processing. Firstly, the present state of the development of optical recording media sets the limits of recording time, light sensitivity, etc.. Secondly, it is impossible to principally presume new optical phenomena which can open unhabituated possibilities for optical processing and optical memory systems. Therefore, I will confine myself to some remarks only.

Presently, the data storage density in optical discs is close to the density of electronic modules in integrated circuits (e.g., the 16 megabit chip) [6.13]. Both of them have reached the physical limit of integration. Further progress in increasing the storage density in optical recording media and the integration level in microelectronic devices is possible only with molecular and nano-structure semiconductor electronics [6.13,16]. The aim of molecular electronics is to simulate biological coding systems (e.g., nucleic acid base-pairs and neural synaptic junctions) in compounds of high molecular mass. Such complex organic compounds are characterized, not only by a high storage density and data transfer rate, but, also are capable of associative information processing. The storage density of such molecular systems is determined by the processor size that, which is close to the physical limit of the "device", i.e., a molecule of approximately 10 nm diameter. Very high data transfer rates may be achieved in molecular systems using parallel data processing. Considering the human eye, for example, where a high data processing rate from the retina to the brain, achieved by one million neurons! A further unique property of biological molecules is their ability of self-assembly and self-repair. The plasticity of molecular circuits might allow high fault tolerance. This may provide the possibility for learning. Finally, the high sensitivity of biochemical systems (e.g., the human eye can detect even one light quantum) is an additional attraction. However, at the present time such synthetic biomolecular systems are far from being a technical memory device. There remain numerous problems in data recording and readout, in particular, the communication of macro molecules with a technical memory device.

Another point of comparison with optoelectronic devices is the data transfer rate. The absolute time limit of optical processes is determined by the uncertainty relation $\Delta E \cdot \Delta t \geq h$. For the visible spectral region ($\Delta E \approx 2$ eV) this gives a minimum time resolution of: $\Delta t \approx 3 \cdot 10^{-16}$ s $= 0.3$ fs, which is close to the shortest generated laser pulse. The time limit, constrained by the uncertainty relation, confines the data rate also. At present, the shortest optical recording time (approximately 100 fs) was reached by holographic recording [6.17]. However, in any optical memory systems there are electronic elements, where the charge transfer time is usually longer than 0.3 fs, which leads to an additional limiting data rate .

It is not easy to combine high light sensitivity, short recording time and long storage time in a single recording material. Further progress in information recording using pico- and femtosecond light pulses and higher storage densities is an important problem for basic research in optical recording and optical, including holographic, processing.

References

Chapter 1

1.1 P.J. van Heerden: Appl. Optics **2**, 392–400 (1963)

1.2 L.E. Hargrove, R.L. Fork, M.A. Pollack: Appl. Phys. Lett **5**, 4–5 (1964)

1.3 M.Di Domenico (Jr), J.E. Geusic, H.M. Marcos, R.G. Smith: Appl. Phys. Lett. **8**, 180–183 (1966)

1.4 D.J. Bradley, A.J.F. Durrant: Phys. Lett. **27A**, 73–74 (1968)

1.5 R.L. Fork, C.H. Brito Cruz, P.C. Becker, C.V. Shank: Opt. Lett. **12**, 483–485 (1987)

1.6 D.J. Amit: *Modelling Brain Function – The World of Attractor Neural Networks* (Cambridge Univ. Press, Cambridge 1989)

1.7 P.H. Lindsay; D.A. Norman: *Human Information Processing* (Acadimic, New York 1972)

1.8 A.B. Marchant: *Optical Recording* (Addison–Wesley Publishing Company, Reading, Massachusetts 1990)

1.9 D.J. Gravesteijn, C.J. van der Poel, P.M.L.O. Scholte, C.M.J. van Uijen: Philips Tech. Rev. **44** (8–10), 250–258 (1989)

1.10 D. Gabor: Nature **161**; 774 (1948)

1.11 D. Casasent (ed): *Optical Data Processing* (Springer, Berlin, Heidelberg 1978) p. 286

1.12 H.J. Caulfield, J. Shamir: Appl. Optics **28** (12), 2184–2186 (1989)

1.13 B.E.A. Salen: *Optical Information Processing and the Human Visual System in Applications of Optical Fourier Transforms* (Academic, New York 1982) pp. 431–463

1.14 A.P. Ginsburg: Nature, **257**, 219–220 (1975)

1.15 R.R.A. Syms: *Practical Volume Holography* (Clarendon Press, Oxford 1990)

1.16 M. Livingston, D. Hubel: Science **240**, 740–749 (1988)

1.17 H.J. Caulfield, J. Kinser, S.K. Rogers: Proc. IEEE **77** (10), 1573–1583 (1989)

1.18 K-Y. Hsu, H-Y. Li, D. Psaltis: Proc. IEEE **78** (10), 1637–1645 (1990)

1.19 R.D. Guenter: *Moderrn Optics* (John Wiley & Sons, New York 1990)

1.20 J.A. Valdamanis, N.H. Abramson: Laser Focus World **27** (2), 111–120 (1991)

1.21 A.W. Lohmann: Phys. Scr. **T23**, 271–274 (1988)

1.22 F.M. Schellenberg, W. Leuth, G.C. Bjorklund: Appl. Optics **25** (18), 3207–3216 (1986)

1.23   S. Arnold, C.T. Liu, W.B. Whitten, J.M. Ramsey: Optics Lett. **16** (6), 420–422 (1991)

1.24   D.B. Carlin, D.B. Kay (editors): *Optical Data Storage* (SPIE New York, 1992)

1.25   A.A. Jamberdino, W. Niblack (ed.): *Image Storage and Retrieval Systems* (SPIE New York, 1992)

1.26   S. Zeki: Scient. Amer. **267** (3), 43–50 (1990)

Chapter 2

2.1   A. Hellemans, B. Bunch: *The Timetables of Science* (Simon and Schuster, New York 1988)

2.2   K.K. Shvarts, Yu.A. Ekmanis: *Proc. Intern. Conf. Defects in Insulating Crystals*, ed. by V.M. Tuchkevich, K.K. Shvarts (Springer, Berlin, Heidelberg 1981), pp. 363–386

2.3   Ch.B. Lushchik: *Physics of Radiation Effects in Crystals*, ed. by R.A. Johnson, A.N. Orlov (Elsevier, Amsterdam 1986), pp. 473–526

2.4   W.C. Röntgen: Ann. Phys. **41**, 449–498 (1913); ibid **64**, 1–195 (1921)

2.5   P.J. van Heerden: Appl. Optics **2**, 393–400 (1963)

2.6   I. Schneider: Appl. Optics **6** (12), 2197–2198 (1967)

2.7   F.S. Chen, Y.T. La Machia, D.B. Fraser: Appl. Phys. Lett. **13**, 223–225 (1968)

2.8   J.B. Thaxter: Appl. Phys. Lett. **15** (7), 210–212 (1969)

2.9   J.P. Huignard, J.P. Herriau, F. Micheron: Appl. Phys. Lett. **26** (5), 256–258 (1975)

2.10   S.A. Keneman: Appl. Phys. Lett. **19** (6), 205–207 (1971)

2.11   D. Chen, J.E. Ready, R.L. Aagard, G.E. Bernal: Laser Focus **4** (5), 18–20 (1968)

2.12   R.W. Teale, D.W. Temple: Phys. Rev. Lett. **19** (16), 904–905 (1967)

2.13   A. Bloom, R.A. Bartolini, D.L. Ross: Appl. Phys. Lett. **24** (12), 612–614 (1974)

2.14   J. Gaynor, S. Aftergut: J. Appl. Phys. **34** (7), 2102–2103 (1963)

2.15   J.C. Urbach, R.W. Meier: Appl. Optics **5** (4), 666–667 (1966)

2.16   A.A. Gorokhovski, A.A. Kaarli, L.A. Rebane: Pisma JETF **20**, 474–479 (1974) *English translation* in JETP Letters (USA)

2.17   A. Szabo: "Frequency selective optical memory." US Pat. No. 38934420 (1975)

2.18   J. Lindmayer, P. Goldsmith, Ch. Wrigley: Laser Focus World **25** (11), 122–127 (1989)

2.19   R.W. Ditchburn: *Light* (Academic Press, London New York San Francisco, 1976)

2.20   A. Ashkin, G.D. Boyd, J.M. Dziedzic, J.M. Smith, A.A. Ballman, J.J. Levinstein, K. Nassau: Appl. Phys. Lett. **9**, 72–74 (1966)

2.21   G.E. Peterson, A.A. Ballman, P.V. Lenzo, P.M. Bridenbaugh: Appl. Phys. Lett. **5**, 62–64 (1964)

2.22   F.S. Chen: J. Appl. Phys. **40**, 3389–3396 (1969)

2.23   J. Kerr: Phil. Mag. **4**, 337 (1875)

2.24   J. Kerr: Phil. Mag. **5**, 157 (1880)

2.25   W.C. Röntgen: Ann. Phys. Chem. **18**, 213–228 (1883)

2.26   A. Kundt: Ann. Phys. Chem. **18**, 228–233 (1883)

2.27   F. Pockels "Über den Einfluss des elektrostatischen Feldes auf das optische Verhalten piezoelektrischer Kristalle", Göttingen, (1895)

2.28   A.G. Chynoweth: Phys. Rev. **102**, 705–714 (1956)

2.29   G. Mayer, F. Gires: Compt. Rend. **258**, 2039–2042 (1964)

2.30   A.M. Glass, D. von der Linde, T.J. Negran: Appl. Phys. Lett. **25**, 233–235 (1974)

2.31   E. Goldberg: US Pat. No. 1838389 (1931)

2.32   A.B. Marchant: *Optrical Recording* – A Technical Overview (Addison–Wesley Publishing Company, Reading, Massachusetts 1990)

2.33   D. Psaltis, M.A. Neifeld, A. Yamamura, S. Kobayashi: Appl. Optics **29** (14), 2038–2057 (1990)

2.34   H.M. Smith (ed.): *Holographic Recording Materials* (Springer, Berlin, Heidelberg 1977)

2.35   L.E. Ravich: Laser Focus World **25** (3), 115–122 (1989)

2.36   M.P. Petrov, A.V. Khomenko, M.G. Shlyagin, M.V. Krasinkova, V.I. Berezkikh: Ferroelectrics **34** (3), 113–115 (1981)

2.37   B.A. Horwitz, F.J. Corbett: Optical Engineering **17** (4), 353–364 (1978)

2.38   H.J. Caulfield, J. Kinser, S.K. Rogers: Proc. IEEE **77** (10), 1573–1583 (1989)

2.39   K.-Y. Hsu, H.-Y. Li, D. Psaltis: Proc. IEEE **78** (10), 1637–1645 (1990)

2.40   A.W. Lohmann: Physica Scripta **T23**, 271–274 (1988)

2.41   D. Psaltis, M.A. Neifeld, A. Ymamura, S. Kobayashi: Appl. Optics **29** (14), 2038–2057 (1990)

2.42   L.T. Kontnik: Laser Focus World **25** (7), 71–72 (1989)

2.43   D. Casasent (ed.): *Optical Data Processing* (Springer, Berlin, Heidelberg 1978)

2.44   R.J. Collier, C.B. Burckhardt, L.H. Lin: *Optical Holography* (Academic, New York 1971)

2.45   W.T. Cathey: *Optical Information Processing and Holography* (John Wiley & Sons, London 1974)

2.46   M. Françon: *Optical Image Formation and Processing* (Academic Press, New York 1979)

2.47   Yu.N. Denisyuk: Dokl. Akad. Nauk SSSR **144**, 1275–1278 (1962) [in Russian] Physics Abstracts **65** (775), p. 1906, No 1679 (1962)

2.48   S.A. Benton: J. Opt. Soc. Am. **59** (11), 1545–1546 (1969)

2.49   A.O. Ozols: Optics and Spectroscopy, vol. **42** (1), 168–174 (1977) *English translation*: Opt. & Spectrosc. (USSR) **44** (6), 1171–1179 (1978)

2.50   H. Kogelnik: Bell Syst. Techn. J. **48** (9), 2909–2947 (1969)

2.51   A. Cozannet, H. Maitre, J. Fleuret, M. Rousseau: *Optique et Telecommunications.* (Eyrolles, Paris 1981)

2.52   A.A. Friesem, J.L. Walker: Appl. Optics **9** (1), 201–214 (1970)

2.53   H. Kurz: Optica Acta **24** (4), 463–473 (1977)

2.54   E. Krätzig, R. Orlowski, V. Dormann, M. Rosenkranz: Proc. SPIE **164**, 33–37 (1979)

2.55   Wai-Hon Lee: Lasers Optronics **9**, 85–87 (1987)

2.56   Y. Komma, S. Kadowaki, Y. Hori, M. Kato: Applied Optics **29** (34), 5127–5130 (1990)

2.57   M. Ekbergs, M. Larsson, S. Hard, B. Nilsson: Optics Letters **15** (10), 568–569 (1990)

2.58   H. Ming, Y. Wu, J. Xie, T. Nakajima: Applied Optics, **29** (34), 5111–5114 (1990)

2.59   M. Haruna, M. Takahashi, K. Wakahayashi, H. Nishihara: Applied Optics, **29** (34), 5120–5126 (1990)

2.60   H. Hosokawa, T. Yamashita: Applied Optics, **29** (34), 5106–5110 (1990)

2.61   L. Levi: *Applied Optics, vol. 2* (John Wiley & Sons, New York 1980), p. 622–627

2.62   G. Thomas, W. Ophey: Physics World **36** (12), 2–7 (1990)

2.63   D.J. Gravesteijn: Applied Optics **27** (4), 736–738 (1988)

2.64   A.O. Zols, K.K. Shvarts: Kvant. Elektron. **9** (12), 2441–2448 (1982)
        English translation: Sov. J. Quantum Electron. (USA) **12**, 1593–1597 (1982)

Chapter 3

3.1   R. Shankar: *Princles of Quantum Mechanics*. (Plenum, New York 1980)

3.2   J.M. Ziman: *Principles of the Theory of Solids*. (Cambridge University Press, London 1972)

3.3   J.S. Blakemore: *Solid State Physics*. (W.B. Saunders, Philadelphia 1974)

3.4   R.W. Ditchburn: *Light.*, (Academic, London 1976)

3.5   M. Young: *Optics and Lasers*. (Springer, Berlin, Heidelberg 1984)

3.6   S.I. Wavilow, W.L. Lewschin: Z. Phys. B**25** (11), 920–936 (1926)

3.7   S.L. Shapiro (ed.): *Ultrashort Light Pulses*. (Springer, Berlin, Heidelberg 1977), Chap. 2

3.8   D. Kales: Laser Focus World **25** (5), 95–114 (1989)

3.9   J. Zeman: Laser Focus World **25** (8), 69–80 (1989)

3.10   Y. Suematsu, K. Iga, S. Arai: "Advanced Semiconductor Lasers", in Proc. IEEE, **80** (3), 383–397 (1992)

3.11   E. Kapon: "Quantum Wire Lasers", in Proc. IEEEE, **80** (3), 398–410 (1992)

3.12   *Lasers and Optronics. 1990 Buying Guide. Technical Handbook* (Elsevier, Amsterdam 1990)

3.13   R.M. Wood, S.K. Sharma, P. Waite: GEC J. Science Technology **48** (3), 141–151 (1982)

3.14   E.W. van Stryland, M.A. Woodall, H. Vanherzeele, M.J. Soilean: Opt. Lett. **10** (10), 490–492 (1985)

3.15   D. von der Linde, A.M. Glass, K.F. Rodgers: Appl. Phys. Lett. **25** (3), 155–157 (1974)

3.16   A. Ozols: Latvijas Zin. Akad. Vestis **1** (522), 108–120 (1991)

3.17   A. Yariv: *Introduction to Optical Electronics.* (Holt, Rinehart and Winston, New York 1976)

3.18   A. Yariv: *Quantum Electronics.* (John Wiley & Sons, New York 1975)

3.19   F. Pockels: *Über den Einfluss des elektrostatischen Feldes auf das optische Verhalten piezoelektrischer Kristalle.* Göttingen (1895)

3.20   A. Yariv, P. Yeh: *Optical Waves in Crystals.* (John Wiley & Sons, New York 1984)

3.21   S.K. Kurtz, J. Jerphagnon, M.M. Choy: *Nonlinear Dielectric Susceptibilities* (Landolt–Bernstein) New Series, Vol. **11,** ed. by K.-H. Hellwege, (Springer, Berlin, Heidelberg 1979)

3.22   A. Winnacker: *Physik von Maser und Laser.* (B.I. Wissenschaftsverlag, Mannheim 1984)

3.23   R.D. Guenter: *Modern Optics.* (John Wiley & Sons, New York 1990)

3.24   J.A. Valdmanis, J.V. Rudd, N.H. Abramson, E.N. Leith, J.L. Lopez: Laser Focus **27,** 111–120 (1991)

3.25   A.H. Guenter, Z.K. Melver: Laser Focus World **26** (6), 103–113 (1990)

3.26   M. Bass, M.L. Stitch (ed.): *Laser Handbook Vol 5* (North-Holland, Amsterdam 1985) p. 84

3.27   M.J. Weber: *CRC Handbook of Laser Science and Technology* (CRC Press, Boca Raton 1982) pp. 243–246

3.28   A.B. Harvey: *International Competitiveness and Business Techniques in Advanced Optics and Imaging,* ed. by E. Sternberg, A.J. Krisiloff, R.R. Schindler (Proc. SPIE 1617, Bellingham 1992) pp. 50–54

3.29   Y.R. Shen: *The Principles of Nonlinear Optics* (John Wiley and Sons, New York 1984)

Chapter 4

4.1   K.K. Rebane: J. Lumin. **31/32,** 744–750 (1984)

4.2   A.A. Gorokhovskii, A.A. Kaarli, L.A. Rebane: J. Exp. Theor. Phys.Lett. **20** (7), 474–479 (1974) *English translation* in JETP Letters

4.3   A. Szabo: "Frequency Selective Optical Memory", US Pat. No 38934420 (1975)

4.4   S. Musikant: "1989 Trends in Optical Materials"; Laser Focus World **25** (1), 140–142 (1989)

4.5   R.R. Neurgeonkar, G. Rackuljic, G.C. Valey, C. Woods: Appl. Optics **26** (2), 220–224 (1987)

4.6   A.A. Ballman, R.L. Byer, D. Eimerl, R.S. Feigelson, B.J. Feldman, L.S. Goldberg, N. Menyuk, C.L. Tang: Appl. Optics **26** (2), 224–227 (1987)

4.7   F.W. Clinard, L.W. Hobbs: "Radiation Effects in Non-Metals" in *Physics of Radiation Effects in Crystals,* ed. by R.A. Johnson, A.W. Orlov (Elsevier, Amsterdam 1986) pp. 387–472

4.8   I. Schneider, M.E. Gringerich: Appl. Opt. **5** (10), 2428–2431 (1976)
4.9   I. Schneider: Appl. Opt. **6** (12), 2197–2198 (1967)
4.10   A.B. Scott, W.A. Smith, M.A. Thompson: J. Phys. Chem. **57** (8), 757 (1953)
4.11   A.E. Hughes, S.C. Jain: Adv. Phys. **28** (6), 717–828 (1979)
4.12   W.E. Bron, W.K. Heller, B. Welber: "Memory device and method of information handling utilizing charge transfer between rare earth ions", US Pat. No 3452332 (1965)
4.13   A.B. Scott, L.P. Bupp: Phys. Rev. **79**, 341–346 (1950)
4.14   A.A. Shatalov: J. Exp. Phys. **29**, 847–856 (1955) [in Russian]
       English translation: Soviet Physics JETP (New York) **2** (4), 725–732 (1956)
4.15   A. Ozols: Izv. Akad. Nauk Latv. SSR, Ser. fiz. tekhn. nauk **3**, 138–140 (1979) [in Russian]
4.16   K.-E. Peiponen, P. Ketolainen, A. Vaittinen, J. Riissanen; Optics Laser Technol. **16**, 203–205 (1984)
4.17   A. Ozols: Izv. Akad. Nauk Latv. SSR, Ser. fiz. tekhn. nauk **5**, 16–25 (1978) [in Russian]
4.18   G.E. Scrivener, M.R. Tubbs: Opt. Commun. **10** (1), 32–36 (1974)
4.19   P. Günther: Phys. Rep. **93** (4), 199–299 (1982)
4.20   M.E. Lines, A.M. Glass: *Principles and Applications of Ferroelectrics and Related Materials* (Clarendon, Oxford 1977)
4.21   O.F. Schirmer, O. Thiemann, M. Wöhlecke: J. Phys. Chem. Sol. **52**, 185–200 (1991)
4.22   Y. Ohmori, M. Yamaguchi, E. Yoshino, Y. Inuishi: Jpn. J. Appl. Phys. **16** (1), 181–182 (1977)
4.23   T.R. Volk, S.A. Shramchenko: *Third Soviet–Japanese Symp. on Ferroelectricity* (Novosibirsk, 1984) p. 173–175
4.24   F.S. Chen: J. Appl. Phys. **40** (8), 3389–3396 (1969)
4.25   A.M. Glass, D. von der Linde, T.J. Negran: Appl. Phys. Lett. **25** (4), 233–235 (1974)
4.26   H.G. Festl, P. Hertel, E. Krätzig, R. von Baltz: Phys. Stat. Sol. (b) **113**, 157–164 (1982)
4.27   P.A. Augustov, K.K. Shvarts: Appl. Phys. A **29**, 169–172 (1982)
4.28   P.A. Augustov, K.K. Shvarts: Appl. Phys. **21** (2), 191–194 (1980)
4.29   K. Shvarts, A. Ozols, P. Augustov, M. Reinfelde: Ferroelectrics **75**, 231–249 (1987)
4.30   P.A. Augustov, K.K. Shvarts: Appl. Phys. **18** (4), 399–401 (1978)
4.31   A.M. Glass, D. von der Linde: Ferroelectrics **10** (1–4), 163–166 (1976)
4.32   D. von der Linde, A.M. Glass, K.F. Rodgers: Appl. Phys. Lett. **25** (3), 155–157 (1974)
4.33   P. Guenter: Ferroelectrics **24** (1–4), 35–42 (1980)
4.34   P. Günter, F. Micheron: Ferroelectrics **18** (1–3), 27–38 (1978)
4.35   D.L. Staebler, J.J. Amodei: J. Appl. Phys. **43** (3), 1042–1049 (1972)
4.36   R.A. Rupp, K. Kerperin, A. Krumins: Ferroelectrics **90**, 75–82 (1989)

4.37   I.B. Barkan, E.M. Baskin, M.V. Entin: Phys. Stat. Solid **59** (1), K97–K102 (1980)

4.38   I.B. Barkan, M.V. Entin, S.I. Marennikov: Phys. Stat. Solid **44** (1), K91–K94 (1977)

4.39   J.B. Thaxter: Appl. Phys. Lett. **15** (7), 210–212 (1969)

4.40   R.A. Rupp, A.E. Krumins, K. Kerperin, R. Matull: Phys. Rev. B **39** (13), 9541–9554 (1989)

4.41   W.D. Johnston: J. Appl. Phys. **41** (8), 3279–3285 (1970)

4.42   E. Krätzig, F. Welz, R. Orlowski, V. Doormann, M. Rosenkranz: Solid State Commun. **34** (10), 817–819 (1980)

4.43   E. Krätzig, R. Orlowsky: Appl. Phys. **15** (2), 133–139 (1978)

4.44   L.A. Boatner, E. Krätzig, R. Orlowski: Ferroelectrics **27** (1–4), 247–250 (1980)

4.45   K. Megumi, H. Kozuka, M. Kobayshi, Y. Furwahta: Appl. Phys. Lett. **30** (12), 631–633 (1977)

4.46   J.P. Huignard, F. Micheron: Appl. Phys. Lett. **29** (9), 591–593 (1976)

4.47   V.M. Fridkin, B.N. Popov, K.A. Verkhovskaya: Appl. Phys. **16** (3), 313–315 (1978)

4.48   R. Orlowski, E. Krätzig: Acta Phys. Austriaca, suppl. **20**, 241–255 (1979)

4.49   J.J. Amodei, D.L. Staebler: Appl. Phys. Lett **18** (12), 540–542 (1971)

4.50   J.J. Amodei, D.L. Staebler, A.W. Stephens: Appl. Phys. Lett **18** (11), 507–509 (1971)

4.51   H. Vormann, E. Krätzig: Solid State Commun. **49** (9), 843–847 (1984)

4.52   J.P. Huignard, J.P. Herriau, F. Micheron: Appl. Phys. Lett. **26** (5), 256–258 (1975)

4.53   M.P. Petrov, S.I. Stepanov, A.V. Khomenko: *Elecrto-optic Recording Materials for Holography and Optical Processing* (Nauka, Moscow 1983) [in Russian]

4.54   D. Casassent (ed.): *Optical Data Processing* (Springer, Berlin, Heidelberg 1978)

4.55   R. Barton, Ch. R. Davis, K. Rubin, G. Lim: Appl. Phys. Lett. **48**, 1255–1257 (1986)

4.56   J. Teteris: Phys. Stat. Solid (a) **83**, K47–K50 (1984)

4.57   M.T. Kostyshin, E.V. Mikhailovskaya, P.F. Romanenko: Fiz. Tverd. Tela **8** (2), 571–572 (1966) [in Russian]

4.58   N. Koshino, K. Utsumi, Y. Goto: Fujitsu Sci. Techn. J. **24** (1), 60–69 (1988)

4.59   G.M. Blom: J. Appl. Phys. **54** (11), 6175–6182 (1983)

4.60   V.V. Kaminskii, Yu.F. Solomonov, V.E. Yegorov, B.I. Smirnov, A.I. Smirnov: Fiz. Tverd. Tela **17** (5), 1546–1548 (1975)
       English translation: Sov. Phys. – Solid State (USA) **17** (5), p. 1015–1018 (1975)

4.61   N.A. Goryunova, B.T. Kolomiets: Izv. AN SSSR. Ser. fiz. **20** (12), 1496–1500 (1956) [in Russian]

4.62   N.F. Mott, E.A. Davis: *Electronic Processes in Non-Crystalline Materials* (Clarendon, Oxford 1979)

4.63   S.R. Elliot: *Amorphous Materials* (Longman Scientific & Technical, London 1990)

4.64   W. Hayes, A.M. Stineham: *Defects and Defect Processes in Non-Metallic Solids* (Wiley, New York 1985) pp. 385–441

4.65   A.C. Wright, G.A.N. Connell, J.W. Allen: J. Non-Cryst. Solids **42** (1–3), 69–86 (1980)

4.66   *Physics on non-tetrahedrally bonded elements and binary compounds*, ed. by O. Madelung, Landolt-Börnstein: Numerical Data and Functional Relationships in Science and Technology. 3. Crystal and Solid State Physics" Vol. 17 (Springer, Berlin, Heidelberg 1983)

4.67   A.C. Wright, A.J. Leadbetter: Phys. Chem. Glasses 17 (5), 122–145 (1976)

4.68   L. Cervinka, A. Hruby: J. Non-Cryst. Sol. 48, 231–264 (1982)

4.69   M. Kastner: Phys. Rev. Lett. 28 (6), 335–357 (1972)

4.70   S.R. Elliot: Adv. Phys. 36 (2), 135–218 (1987)

4.71   A.E. Owen: "Electron Transport in Chalcogenide Glasses", in *Coherence and Energy Transfer in Glasses*, ed. by P.A. Fleury, B. Golding (Plenum, New York 1984) pp. 243–278

4.72   *Handbook of Chemistry and Physics* (CRC, New York 1983–1984)

4.73   N.F. Mott, E.A. Davis, R.A. Street: Phil. Mag. 32, 961–996 (1975)

4.74   M. Kastner, D. Adler, H. Fritzsche: Phys. Rev. Lett. 37 (22), 1504–1507 (1976)

4.75   M. Klinger: Phys. Reports 165 (56), 275–397 (1988)

4.76   J.P. de Neufville, S.C. Moss, S.R. Ovshinsky: J. Non-Cryst. Solids 13 (2), 191–223 (1974)

4.77   K. Tanaka: J. Non-Cryst. Solids 35/36, 1023–1034 (1980)

4.78   K. Tanaka: Thin Solid Films 157, 35–41 (1988)

4.79   G. Pfeiffer, M.A. Paesler, S.C. Agarwal: J. Non-Cryst. Solids 130, 111–143 (1991)

4.80   M. Yamada, Y. Ohmori, K. Takada, M. Kobayashi: Appl. Optics 30 (6), 682–688 (1991)

4.81   S.G. Bishop, U. Strom, P.C. Taylor: Phys. Rev. B 15, 2278–2294 (1977)

4.82   S.A. Kenemann: Thin Solid Films 21 (2), 281–285 (1974)

4.83   K. Petkov: Thin Solid Films 21 (2), 281–285 (1974)

4.84   M. Mitkova, Z. Boncheva-Mladenova: J. NonCryst. Solids 90, 589–592 (1987)

4.85   V.M. Lyubin: "Photographic Processes in Glassy Chalcogenide Semiconductors", in *Non-Silver Halide Photographic Processes*, ed. by A.L. Kartuzhanskii (Khimiya, Leningrad 1984) pp. 193–222 [in Russian]

4.86   P. Stradins, K. Shvarts, J. Teteris: J. Non-Cryst. Solids 114, 79–81 (1989)

4.87   A. Hamou, G. Fleury, C. Viger: Thin Solid Films 123 (1), 87–92 (1985)

4.88   N. Sakai, T. Kajiwara: Jpn. J. Appl. Phys. 21, 1383–1393 (1982)

4.89   K.K. Shvarts, Yu. N. Shunin, J.A. Teteris: Cryst. Latt. Def. Amorph. Mat. 17, 133–138 (1987)

4.90   I. Manika, J. Teteris,: Phys. Stat. Solidi (a) 80, K121–K123 (1983)

4.91   K.Shvarts, J. Teteris, I. Manika, M. Reinfelde, V. Gerbreder: J. Non-Cryst. Solids 90, 509–512 (1987)

4.92   K.K. Shvarts, M.J. Reinfelde, A.O. Ozols: Proc. Physics Institute Estonian Academy of Sciences 63, 107–118 (1989) [in Russian]

4.93   P. Ketolainen, A. Ozols, V. Pashkevich, M. Reinfelde, O. Salminen, K. Shvarts: Avtometrija 4, 103–106 (1988)

4.94   M. Poup, Ch. Swenberg: *Electronic Processes in Organic Crystals* (Oxford University Press, London 1982)

4.95   E.A. Silinsh: *Organic Molecular Crystals: Their Electronic States*, in Springer Ser. Solid State Sci., Vol. **10** (Springer, Berlin, Heidelberg 1980)

4.96   G. Thomas, W. Ophey: Physics World **36** (12), 2–7 (1990)

4.97   S. Wu, J. Chen, P. Low, F. Lin: "Randomly Addressable Read-Write-Erase Holographic Memory Systems Based on a Dye-Polymer Recording Medium", in *Image Storage and Retrieval Systems*, ed. by A.A. Jamberdino, W. Niblack (Proc. SPIE 1662, Washington 1992) pp. 168–174

4.98   R.B. Gross, K. Can Izgi, R.R. Birge: "Holographic Thin Films, Spatial Light Modulators and Optical Associative Memories Based on Bacteriorhodopsin" (Proc. SPIE 1662, Washington 1992) pp. 186–196

4.99   A.S. Dvornikov, P.M. Rentzepis: "Studies on 3-D Volume Memory" (Proc. SPIE 1662, Washington 1992) pp. 197–204

4.100   R. Thoma, N. Hampp, C. Brauchle: Opt. Lett. **16** (9), 651–653 (1991)

4.101   R.R. Birge, P.A. Fleitz, R.B. Gross, J.C. Izgi, A.F. Lawrence, J.A. Stuart, J.R. Tallent: Proc. IEEE MEBC **12** (4), 1788–1789 (1990)

4.102   R. Kachru, Y. Sheng Bai, X.-A. Shen, D.L. Huestis: "Random Access Stimulated Echo Optical Cache Memory", in *Image Storage and Retrieval Systems*, ed. by A:A: Jamberdino, W. Niblack (Proc. SPIE 1662, Washington 1992) pp. 205–210

4.103   R.A. Bartolini: J. Vac. Sci. Technol. **18** (1), 70–74 (1981)

4.104   R.R. Birge: Biochem. Biophys. Acta **1016**, 293–327 (1990)

4.105   F.V. Bunkin, N.N. Vsevolodov, A.B. Druzhko, B.I. Mitsner, A.M. Prokhorov, V.V. Savranskii, N.W. Tkachenko, T.B. Shevchenko: Sov. Tech. Phys. Lett. **7**, 630–631 (1981)

4.106   V.A. Poltoratskii, N.N. Vsevolodov: Zh. Tech. Fiz. (USSR) **55** (10), 2093–2094 (1985) *English translation*: Sov. Phys. – Tech. Phys. (USA) **30** (10), p. 1235 (1985)

4.107   N. Hampp, C. Brauchle, D. Oesterhelt: Biophys. J. **58**, 83–93 (1990)

4.108   M. Yoshida, K. Ohno, Y. Takeuchi, Y. Kagawa: Biochem. Biophys. Res. Comm. **75**, 1111–1116 (1977)

4.109   M. Nakasako, M. Kataoka, F. Tokunaga: FEBS. Letts. **254**, 211–214 (1989)

4.110   R.R. Birge: Annu. Rev. Phys. Chem. **41**, 683–733 (1990)

4.111   G.I. Lashkov: "The Volume Recording Material Reoxan", in *Non-Silver Halide Photographic Processes*, ed. by A.L. Kartuzhanskii (Khimiya, Leningrad 1984) pp. 130–145 [in Russian]

4.112   H.M. Smith (ed.): *Holographic Recording Materias* (Springer, Berlin, Heidelberg 1977)

4.113   F.M. Schellenberg, W. Lenth, G.C. Bjorklund: Appl. Optics **25** (18), 3207–3216 (1986)

4.114 L.A. Rebane, A. Gorokhovski, R. Kaarli: Opt. Commun. **16**, 282–284 (1976)

4.115 K.K. Rebane, L.A. Rebane: "2. Basic Principles and Methods of Persistent Spectral Hole-Burning", in *Persistent Spectral Hole-Burning: Science and Applications*, ed. by W.E. Moerner, Topics Curr. Phys., Vol. 44 (Springer, Berlin, Heidelberg 1988) pp. 17–77

4.116 A. Winnacker, M. Shelby, R.M. Macfarlane: Opt. Lett. **10**, 350–353 (1985)

4.117 S. Arnold, C. Liu, W.B. Whiten, J.M. Ramsey: Opt. Lett. **16** (6), 420–422 (1991)

4.118 T.W. Mossberg: Opt. Lett. **7** (2), 77–79 (1982)

4.119 A.K. Rebane, R. Kaarli, P. Saari: Zh. Eksp. Teor. Fiz. **38**, 320–322 (1983)
       English translation: JETP Lett. **38**, 383–385 (1983)

4.120 A.K. Rebane, R.K. Kaarli, P.M. Saari: J. Mol. Struc. **114**, 343–345 (1984)

4.121 P.M. Saari, R.K. Kaarli, A.K. Rebane: J. Opt. Soc. Am. B**3**, 527–533 (1986)

4.122 A. Rebane, S. Bernet, A. Renn, U.P. Wild: Optics Comm. **86**, 7–13 (1991)

4.123 U.P. Wild, A. Renn, C. De Caro, S. Bernet: Appl. Optics **29**, 4329–4334 (1990)

4.124 K. Ishii, T. Takeda, K. Itao, R. Kaneko: "Mass Storage Technology in Networks", in *Image Storage and Retrieval Systems*, ed. by A.A. Jamberdino, W. Niblack (Proc. SPIE 1662, Washington 1992) pp. 2–9

4.125 A. Rebane, J. Feinberg: Letters to Nature **351**, 378–380 (1991)

4.126 S. Bernet, B. Kohler, A. Rebane, A. Renn, U.P. Wild: J. Opt. Soc. America B**9** (6), 987–991 (1991)

4.127 V. Lyubin, V.K. Tikhomirov: J. Non-Cryst. Solids **135**; 37–48 (1991)

Chapter 5

5.1   A.B. Marchant: *Optical Recording – A Technical Overview* (Addison – Wesley, Reading, Massachusetts 1990)

5.2   L.E. Ravich: Laser Focus World **25** (5), 115–122 (1989)

5.3   A.E. Bell: Proc. SPIE **382**, 2–15 (1983)

5.4   K.A. Rubin, M. Chen: Thin Solid Films 181, 129–139 (1989)

5.5   K. Torazawa, S. Sumi, S. Murata, S. Minechika, Y. Ishi: Appl. Opt. **25** (22), 3990–3995 (1986)

5.6   Y. Sako, T. Suzuki: Appl. Opt. **25** (22), 3996–4000 (1986)

5.7   B.A.J. Jacobs: Vacuum **35** (10–11), 445–446 (1985)

5.8   R.A. Bartolini: J. Vac. Sci. Techn. **18** (1), 70–74 (1981)

5.9   G. Thomas, W. Ophey: Physics World **36** (12), 2–7 (1990)

5.10   D.J. Gravesteijn, C. van de Poel, P.M.L. Scholte, C.M.J.van Uijen: Philips Tech. Rev. **44** (8/9), 250–258 (1989)

5.11   Y. Kimura, S. Sugama, Y. Ono: Appl. Opt. **27** (4), 668–671

5.12   D. Psaltis, M.A. Neifeld, A. Yamamura, S. Kobayashi: Appl. Optics **29** (4), 2038–2057

5.13   D.J. Gravesteijn, J. van der Veen: Philips Tech. Rev. **41** (11/12), 325–333

5.14   S.R. Ovshinsky: IEEE Trans. *ED–20*, 91–95 (1973)

5.15   J. Feinleib, J. De Neufville, S.C. Moss, S.R. Ovshinsky: Appl. Phys. Lett. **18** (6), 254–257 (1971)

5.16   R.J. von Gutfeld: Appl. Phys. Lett. **22**, 257–259 (1973)

5.17   K. Weiser, R.J. Gambino, J.A. Reinhold: Appl. Phys. Lett. **22**, 48–50 (1973)

5.18   M. Chen, V. Marrello, U.G. Gerber: Appl. Phys. Lett. **41** (9), 894–896 (1982)

5.19   G.M. Blom: J. Appl. Phys. **54** (11), 6175–6182 (1983)

5.20   S.R. Herd, K.Y. Alin, R.J. von Gutfeld et al.: J. Appl. Phys. **53** (5), 3520–3522 (1982)

5.21   D.Y. Lou, M. Blom, G.C. Kenney: J. Vac. Sci. Tech. **18** (1), 78–86 (1981)

5.22   P.C. Clemens: Appl. Opt. **22**, 3165–3167 (1983)

5.23   M. Takenaga, N. Yamada, K. Nishiuchi et al.: J. Appl. Phys. **54** (9), 5376–5380 (1983)

5.24   W.Y. Lee, H. Coufal, C.R. Davis, V. Jipson, G. Lim, W. Parrish, F. Sequeda: J. Vac. Sci. Tech. A**4**, 2988–2993 (1986)

5.25   M. Chen, K.A. Rubin, R.W. Barton: Appl. Phys. Lett. **49**, 502–504 (1986)

5.26   Y. Sato, Y. Maeda, H. Andoh, I. Ikuta, M. Nagai, N. Tsuboi, H. Minemura, M. Ishigaki: Proc. Soc. Photo-Opt. Inst. Eng. **1078**, 11–14 (1989)

5.27   M. Terao, Y. Miyauchi, K. Andao, H. Yasuoka, R. Tamura: Proc. Soc. Photo-Opt. Inst. Eng. **1078**, 2–6 (1989)

5.28   T. Ochta, M. Uchida, K. Yoshioka, S. Furukawa, K. Kotera: Proc. Soc. Photo-Opt. Inst. Eng., Techn. Digest Ser. **1**, 14–17 (1989)

5.29   Y.S. Tyan, D.R. Preuss, F. Vazan, S.J. Marino: J. Appl. Phys. **59** (3), 716–719 (1986)

5.30   *Memory Devices and Technology*, Electron. Eng. **61** (755), 49–54 (1989)

5.31   E.S. Rotchild: Laser Focus/Electro-Optics **21** (5), 52–58 (1985)

5.32   R. Barton, Ch.R. Davis, K. Rubin, G. Lim: Appl. Phys. Lett. **48**, 1255–1257 (1986)

5.33   P. Stradins, K. Shvarts, J. Teteris: J. Non-Cryst. Solids **114**, 79–81 (1989)

5.34   J.C. Phillips: Phys. Today **35** (2), 27–33 (1982)

5.35   J.C. Phillips: Phys. Rev. B**25** (2), 1397–1400 (1982)

5.36   N. Koshino, K. Utsumi, Y. Goto: Fujitsu Sci. Tech. J. **24** (1), 60–69 (1988)

5.37   V.B. Jibson: "Directions in Optical Storage", in *Optical Data Storage*, ed. by J.J. Burke, T.A. Shull, N. Imamura (Proc. SPIE 1499, Washington 1991) pp. 2–3

5.38   S. Okamine, M. Teraro, K. Andoo, Y. Miyanchi: "Reversible Phase-Change Optical Recording by Using Microcellular GeSbTeCo Recording Film", in *Optical Data Storage*, ed. by J.J. Burke, T.A. Shull, N. Imamura (Proc. SPIE 1499, Washington 1991) pp. 166–170

5.39   E. Ohno, K. Nishinchi, Y. Yamada, N. Akahira: "Erasable Disc Utilizing Phase Change Material and Multi-Pulse Recording Method", in *Optical Data Storage*, ed. J.J. Burke, T.A. Shull, N. Imamura (proc. SPIE 1499, Washington 1991) pp. 171–179

5.40   T. Ohta, S. Furukawa, K. Yoshioka, M. Uchida, K. Inone, T. Akiyama, K. Nagata, S. Nakamura: "Accelerated Aging Studies for Phase Change Type Disc Media", in *Optical Data Storage*, ed. by M. de Haan, Y. Tsunoda (Proc. SPIE 1316, Washington 1990) pp. 367–373

5.41   M. Suzuki, K. Furuya, K. Nishimura,, K. Mori, I. Morimoto: "Disc Structure and Writing Method for high Performance Phase Change Erasable Optical Disc", in *Optical Data Storage*, ed. by M. de Haan, Y. Tsunoda (Proc. SPIE 1316, Washigton 1990) pp. 374–381

5.42   K. Ishii, T. Pakeda, K. Itao, R. Kaneko: "Mass Storage Technology in Networks", in *Storage and Retrieval*

5.43   H.-P.D. Shich, M.H. Kryder: Appl. Phys. Lett. **49** (8), 473–474 (1986)

5.44   M.D. Schultz, M.H. Kryder: IEEE Trans. MAG–**22** (5), 925–927 (1986)

5.45   J. Hecht: Laser and Optronics (9), 77–79 (1987)

5.46   J. Bohm, S. Kusch: J. Inf. Rec. Mater. **14** (4), 235–244 (1986)

5.47   S. Matsushita, K. Sunago, Y. Sakura: Jap. J. Appl. Phys. **15** (4), 713–714 (1976)

5.48   N. Imamura, Y. Mimura, T. Kobayashi: IEEE Trans. MAG–**12** (2), 55–61 (1976)

5.49   T. Chen, G.B. Charlan: IEEE Trans. MAG–**16** (5), 1194–1196 (1980)

5.50   L.M. Holms: Laser Focus/Electro-Optics **24** (8), 24–25 (1987)

5.51   M. Hartman, B.A.J. Jacobs, J.J. Braat: Philips Tech. Rev. **42** (2), 37–47

5.52   W. Saffady: *Optical Discs vs. Magnetic Storage* (Meckler, London 1990)

5.53   S. Ogawa, M. Maeda: "Magento-Optical Recording Enhanced by Magnetic Recording Techniques", in *Storage and Retrieval Systems and Applications*, ed. by D.H. Davies, H.P.D., Shieh (Proc. SPIE 1248, Washington 1990) pp. 28–35

5.54   J. Saito, H. Akasaka: "Direct Overwritable Magneto-Optical Exchange Coupled Multilayered Disc by Laser Power Modulation Recording", in *Optical Data Storage*, ed. by J.J. Burke, T.A. Shull, N. Imamura (Proc. SPIE 1499, Washington 1991) pp. 44–54

5.55   K. Naito, T. Numata, K. Nakashima, M. Maeda, N. Koshino: "Dy FeCo Magneto-Optical Discs With a Ce-Sio_2 Protective Film", in *Optical Data Storage*, ed. by J.J. Burke, T.A. Shull, N. Imamura (Proc. SPIE 1499, Washington 1991) pp. 386–392

5.56   S. Hashimoto, H. Matsuda, Y. Ochiai: "Ultrathin Co/Pt and Co/Pd New Magneto-Optical Recording Media", in *Storage and Retrieval Systems and Applications*, ed. by D.H. Davies, H.P.D. Shieh (Proc. SPIE 1248, Washington 1990) pp. 36–48

5.57   W. Stolz, R. Bernhardt: *Dosimetrie ionisierender Strahlung* (Akademie, Berlin, 1981)

5.58   G.I. Vlasov, R.A. Kalninsh, L.E. Nagli, V.P. Objedicov, I.K. Plyavin, A.K. Tale: Avtometrija **1**, 66–84 (1980) [in Russian]

5.59   J. Lindmayer, P. Goldsmith, Ch. Wrigley: Laser Focus World **25** (11), 122–127 (1989)

5.60   J. Lindmayer: Sensors **3** (3), 37 (1986)

5.61   J. Lindmayer: Solid State Tech. **31** (8), 135 (1988)

5.62   P. Goldsmith, J. Lindmayer, Ch.Y. Wrigley: "Electron Trapping – A New Approach to Rewritable Optical Data Storage", in *Optical Data Storage*, ed. by M. de Haan, Y. Tsunoda (Proc. SPIE, washington 1990), pp. 312–320

5.63   S. Albin, J.D. Satira, L.C. Watkins, T.A. Shull: "An Erasable Optical Memory Using Stimulated Electronic Transitions Concept", in *Optical Data Storage*, ed. by M. de Haan, Y. Tsunoda (Proc. SPIE, Washington 1990), pp. 358–362

5.64   M. Thoms, H. von Seggern, A. Winnacker: Phys. Rev. B **44** (17), 9240–9247 (1991)

5.65   A. Winnacker: "X-ray Imaging with Photostimulated Storage Phosphors and Future Trends" – Physica Medica **9** (2) (1993) – in print

Chapter 6

6.1   A.B. Marchant: *Optical Recording – A Technical Overview* (Addison – Wesley, Reading, Massachusetts 1990)

6.2   J.A. McCormick: *A Guide to Optical Storage Technology* (Dow Jones–Irwin, Homewood, Illinois 1990)

6.3   D. Psaltis, M.A. Neifeld, A. Yamamura, S. Kobayashi: Appl. Optics **29** (14), 2938–2057 (1990)

6.4   S. Hunter, F. Kiamilev, S. Esener, D.A. Phartenopoulos, P.M. Rentzepis: Appl. Optics **29** (14), 2058–2066 (1990)

6.5   S. Wu, J. Chen, P. Low, F. Lin: "Randomly Addressable Read–Write–Erase Holographic Memory systems Based on a Dye–Polymer Recording Medium", in *Image Storage and Retrievable Systems*, ed. by A.A. Jamberdino, W. Niblack (Proc. SPIE 1662, Washington 1992) pp. 168–174

6.6   K.K. Rebane, L.A. Rebane: "2. Basic Principles and Methods of Persistent Spectral Hole–Burning", in *Persistent Spectral Hole–Burning: Science and Applications*, ed. by W.E. Moerner, Topics Curr. Phys., Vol. 44 (Springer, Berlin, Heidelberg 1988) pp. 17–77

6.7   S. Bernet, B. Kohler, A. Rebane, A. Renn, V.P. Wild: J. Opt. Soc. Am. B**9** (6), 987–991 (1992)

6.8   F.M. Schellenberg, W. Lenth, G.C. Bjorklund: Appl Optics **25** (18), 3207–3216 (1986)

6.9   R. Kachru, Yu.S. Rai, Xiao–An Shen, D.L. Huestis: "Random Access Stimulated Echo Optical Cache memory", in *Image Storage and Retrivial Systems*, ed. by A.A. Jamberdino, W. Niblack (Proc. SPIE 1662, Washington 1992) pp. 205–210

6.10   S. Arnold, C.T. Liu, W.B. Whitten, J.M. Ramsey: Optics Lett. **16** (6), 420–422 (1991)

6.11   J.A. Oko, J.C. Wolfe: J. Vac. Sci. Technol. B**5** (1), 102–109 (1987)

6.12   K. Ishii, T. Takeda, K. Itao, R. Kaneko: "Mass Storage Technology in Networks", in *Image storage and Retrivial Systems*, ed. by A.A. Jemberdino, W. Niblack (Proc. SPIE 1662, Washington 1992) pp. 2–9

6.13   A. Chiabrera, E. Di Zitti, F. Costa, G.M. Bisio: J. Phys. D Appl. Phys. **22**, 1571–1579 (1989)

6.14   V.B. Jipson: "Directions in Optical Storage", in *Optical Data Storage*, ed. by J.J. Burke, T.A. Shull, N. Imamura (Proc. SPIE 1499, Washington 1991), pp. 2–3

6.15   D. Chemla: Physics Today **6**, 22–23 (1993)

6.16   D. Chemla: Physics Today **6**, 46–52 (1993)

6.17   H. Chen, Y. Chen, D. Dilworth, E. Leith, J. Lopez, J. Valdmanis: Optics Lett **16** (7), 487–489 (1991)

Subject Index

Recording Medium Index

Printing: Mercedesdruck, Berlin
Binding: Buchbinderei Helm, Berlin